W9-CGN-747

ALSO BY JAY INGRAM

The Science of Why, Volume 3: Answers to Questions About Science Myths, Mysteries, and Marvels

The Science of Why^2: Answers to Questions About the Universe, the Unknown, and Ourselves

The Science of Why: Answers to Questions About the World Around Us

The End of Memory: A Natural History of Aging and Alzheimer's

Fatal Flaws: How a Misfolded Protein Baffled Scientists and Changed the Way We Look at the Brain

Theatre of the Mind: Raising the Curtain on Consciousness

Daily Planet: The Ultimate Book of Everyday Science

The Daily Planet Book of Cool Ideas: Global Warming and What People Are Doing About It

The Science of Everyday Life

The Velocity of Honey: And More Science of Everyday Life

The Barmaid's Brain and Other Strange Tales from Science

The Burning House: Unlocking the Mysteries of the Brain

A Kid's Guide to the Brain

Talk Talk Talk: Decoding the Mysteries of Speech

It's All in Your Brain

Real Live Science: Top Scientists Present Amazing Activities Any Kid Can Do

Amazing Investigations: Twins

Why Do Onions Make Me Cry?

Answers to Everyday Science Questions You've Always Wanted to Ask

Jay Ingram

Published by Simon & Schuster

NEW YORK LONDON TORONTO SYDNEY NEW DELHI

Simon & Schuster Canada
A Division of Simon & Schuster, Inc.
166 King Street East, Suite 300
Toronto, Ontario M5A 1J3

This Simon & Schuster Canada edition April 2019

SIMON & SCHUSTER CANADA and colophon are trademarks of Simon & Schuster, Inc.

For information about special discounts for bulk purchases, please contact Simon & Schuster Special Sales at 1-800-268-3216 or CustomerService@simonandschuster.ca.

Illustrations by Tony Hanyk, tonyhanyk.com

Manufactured in the United States of America

1 3 5 7 9 10 8 6 4 2

Library and Archives Canada Cataloguing in Publication

Ingram, Jay, author
Why do onions make me cry? / Jay Ingram.
Issued in print and electronic formats.
ISBN 978-1-982110-83-3 (hardcover).—ISBN 978-1-982110-84-0 (ebook)
1. Science—Popular works. 2. Science—Miscellanea. I. Title.
Q162.I56 2019 500 C2018-903428-9
C2018-903429-7

ISBN 978-1-9821-1083-3
ISBN 978-1-9821-1084-0 (ebook)

Contents

Part 3: The Animal Kingdom

Part 4: The Natural World

Part 5: Weird Science & Machines

Why Do Onions Make Me Cry?

Part 1
The Human Body

Why do onions make me cry?

The Bible describes hell as a giant lake of "fire and brimstone." Brimstone is really just an ancient term for the element sulfur. A lake of brimstone would have a pretty awesomely powerful smell, but you don't have to wait to experience it—look no further than the small but potent onion.

As you cut an onion, the knife blade breaks down the tissues in it, releasing chemicals that normally never come in contact with each other. When they do come into contact, they begin to rearrange themselves into new combinations of sulfur-containing compounds. Ultimately, a molecule named syn-propanethial-S-oxide is released into the air. When receptors in the cornea of your eye sense its presence, they release protective tears to wash it away.

Did You Know . . . When you cut garlic, the same process of breaking down and re-forming molecules occurs . . . but we don't cry when we cut garlic because the chemical pathways triggered by the cutting don't create compounds that irritate our eyes.

That tear reaction peaks roughly half a minute after you first cut into the onion, and it takes about five minutes to go away. It's impossible to turn off this chemical reaction, but there are some ways to make the process a little less painful. Some people suggest putting the onion in the fridge or freezer before you cut it. Chemical reactions are slower at lower temperatures, so if the onion is cold when you first cut into it, you might be able to finish the job before the tear-inducing chemicals are released. There have been all sorts of other suggestions, such as cutting the onion while it's under water, covering the lower part of your face with a paper towel, running a fan, dicing the onion under a running stove vent, or even wearing safety goggles. So far, nothing has proved absolutely foolproof—the onion always wins.

Can humans echolocate?

The short answer is yes, and there are individuals who prove it. Daniel Kish has been blind since the age of thirteen months, but now, as an adult, he can camp in the woods, swim, dance, and otherwise navigate life, all thanks to his extraordinary echolocation skills. Other blind echolocators can navigate around a city by skateboard or bicycle. And yet, even with this evidence staring us in the face, it seems unbelievable that humans can echolocate. Most of us have never had the slightest hint that we might be able to do so. But if you pause for a moment, you can see many everyday examples of echolocation in practice—a contractor knocking on a wall to determine where the studs are, or a doctor tapping a patient's abdomen to assess the health of the organs within.

This general lack of awareness is the reason that the first investigators who explored the mechanics of echolocation came up with some weird explanations. Some believed that echolocation was made possible by changes in air pressure on the eardrums or the skin of the face (an early term for it was "facial vision"). Others attributed the cause to magnetism, electricity, or the ether. One expert even argued that touch receptors in the skin were actually tiny eyes, and at one point, the subconscious was thought to be the source of the ability. Although those ideas were inventive, they lacked the benefit of experimental evidence. Finally,

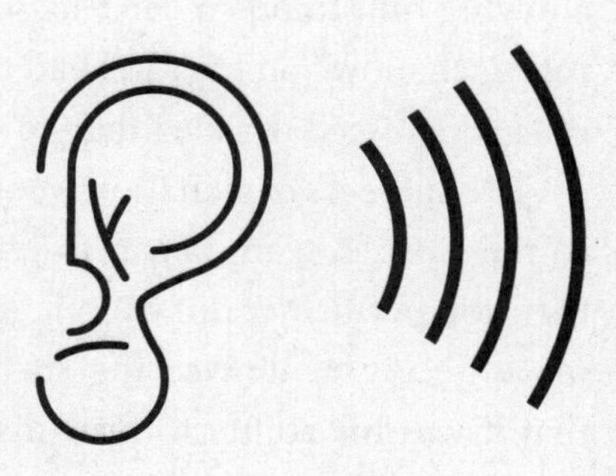

around the turn of the twentieth century, Theodor Heller, a German scientist, actually tested blind people's abilities to sense the presence of an obstacle in front of them. He declared that his subjects were able to pick up changes in the sounds of their footsteps from nine to twelve feet (three to four meters) away from the obstruction, and so, he concluded, acoustics provided the crucial signals that allowed people to navigate without seeing.

The first echolocation experiments were conducted at Cornell University in the early 1940s. The researchers had their subjects—two blind and two sighted—walk down a long hallway that was either empty or had a large sheet of Masonite randomly placed in it. To make the experience as consistent as possible, each subject was blindfolded, and the experiments were run either on a bare wooden floor with the subjects wearing shoes or on a carpet with the subjects wearing socks. When the subjects felt that they were approaching either the end wall or the Masonite board, they were to raise their right hand when they first became aware of the obstacle and then raise their left hand when they felt they were about to bump into it.

Striking results emerged from these tests. As you'd expect, the blind participants fared better than the sighted ones at first: the sighted subjects ran into the wall more than a dozen times during the first trials and often veered so far sideways that they risked running into the side walls of the hall. The top performer initially was one of the blind subjects who was consistently able to detect the wall from more than twenty feet (six meters) away. When the testers asked him how he could sense the obstacles, he explained that somehow the sensitivity of his face was allowing him to do so. He was so convinced of this that he said the wall cast a "shadow" on his forehead that he could feel, even if he couldn't see it. He dismissed the idea that he was actually using his hearing.

The subject's explanation went against the scientists' expectations, and so the Cornell team began to refine the experiment to test how exactly their top performer did so well. After the subject performed dramatically more poorly while wearing socks on the carpet, the scientists realized that it was his ability to hear his footfalls, not some phantom presence felt on his face that was the key to his success. To ensure that sound, not touch, was the important factor, the testers ran another version of the experiment in which they enclosed the man's arms and hands in leather gauntlets and covered his head with a hood that extended over his chest.

With all that equipment on, the subject could still hear, but he was unable to detect a breeze from an electric fan placed ten feet (three meters) away. Even outfitted like that, he could still sense the presence of an object in his path.

With that new information in hand, the scientists ran the tests with all the subjects wearing earplugs. They all performed less well, proving that hearing was crucial to echolocation, for both blind and sighted subjects. The more trials the scientists ran and the more they refined the experiment, the better the sighted group became at echolocation. By the end, one of the sighted subjects actually performed better than one of the two blind ones.

These initial experiments were followed up by variations of one kind or another in various labs, all of which supported the idea that echolocation is a hearing sense, not an air pressure sense. Later experiments also started to flesh out the details of this hearing sense, such as the fact that echolocation relies on the pitch of the sound issued, not its volume. In one study, subjects were encouraged to make whatever noises they felt would best reveal their environment. Instead of just relying on their footsteps, the subjects made clicking noises, snapped their fingers, hissed, whistled, and even sang the "do, re, mi" scale. The common element of the most effective strategies was that the subjects exploited the higher-frequency sound range.

Another study showed that echolocation was more successful when the subjects were able to move their heads from side to side. Subjects were trained to echolocate their way down a hall, then they tried to do the same in a video simulation of the same hall, using the same echolocation sounds. In one version they were allowed to move their heads and bodies to align themselves with the center of the hall; in the other, they had to remain still and adjust their heads with a joystick. The joystick part of the experiment was a failure—the subjects consistently ran into the walls, showing that head movements are crucial to being a good echolocator, probably because of the slight differences in patterns of echoes depending on where the ears are pointing.

TRY THIS: Hold your hand, palm facing you, at arm's length in front of your face. Begin hissing. Slowly move your hand closer to your mouth. You will notice differences in the sound. It changes as a result of the outgoing hiss interfering with the returning echo. It might be difficult for you to attribute this directly to the echoes (just as it was difficult for one of the subjects in the Cornell experiments to believe he was using sound and not pressure on his face), but there's no mistaking the fact that something is changing.

One of the most fascinating discoveries about echolocation has been the modern revelation of what goes on in the brain while it's happening. Mel Goodale and his colleagues at the University of Western Ontario performed an experiment in which they placed two adept echolocators—blind individuals who were able to play basketball or to mountain bike—into an MRI and mapped the areas of their brains that were activated by echoes. They found that the activity in auditory parts of the subjects' brains did not ramp up when hearing echoes, but the activity in the parts of the brain responsible for visual information did. As these individuals began to rely increasingly on echolocation for their perception of the world around them, they apparently abandoned the auditory parts of the brain and recruited the visual. In the case of these particular individuals, the one who had become blind more recently did not show as strong a response in the visual areas of the brain, suggesting that it takes time to reroute this information.

You'd be hard-pressed to match a subtle and sophisticated piece of biology like this with a machine. What's more, humans have proven themselves able to train and improve that ability, suggesting that it's a skill within everyone's reach, regardless of their dependence on sight. What's more, those who echolocate are not limited to simply localizing objects, but can also make accurate judgments about shape and motion. Experienced human echolocators can determine the difference between two objects only 1 inch (2.5 centimeters) apart located 3 feet (1 meter) away, and some can even differentiate by echo the difference in texture between velvet and denim. Contrast that with the sad fact that sonar operators in submarines have often failed to discriminate between a whale and an enemy vessel and as a result depth-charged the whale. Most of us aren't aware we have these tools built into us, but that's simply because we've never had any reason to think about them or to try them out.

Of course, no matter how talented some humans might be at echolocation, bats are the gold standard in the field (closely followed by dolphins and whales). The discoveries that led to our recognition of bats' mastery of biological sonar were made by a curious collection of people. First, Lazzaro Spallanzani—better known for being the first person to perform in vitro fertilization (in frogs)—demonstrated in 1790 that bats can navigate without being able to see. Unfortunately, he proved that by putting out their eyes and showing that the loss of their eyesight wasn't a hindrance to their ability to fly.

Next, Hiram Maxim, the inventor of the machine gun, proposed that bats emit sounds below the range of human hearing and listen for the returning echoes. He didn't test his idea with experiments, but he was correct on one count—we can't hear bats' sounds—and wrong on another—their calls are actually above the range of human hearing. (After the *Titanic* sank in 1912, Maxim suggested that ships could use echolocation to detect icebergs at long range.)

It was eventually established that bats use ultra-high-frequency sounds to detect and capture their insect prey. That sounds easy, but it most definitely is not. Consider that a bat hunting near trees, bushes, and buildings must be able to distinguish echoes returning from those objects. And it must be able to distinguish this noise "clutter" from the wingbeats of a flying insect. Bats do that by changing the frequency of their clicks. When a bat is scanning its environment, it emits about ten to twenty clicks every second. But when it picks up an echo returned from a target, it increases that rate to as many as two hundred clicks a second in order to zero in on the position and identity of its prey.

Some species of moth, upon hearing a bat's clicks, will instantly fold their wings and drop straight down to avoid getting caught. Bats, in turn, have evolved ways of adjusting their clicks to take in a wide area in which even a plummeting bug will be tracked. Some moths have counter-adapted and evolved long extensions to the backs of their wings that twirl in flight to attract bats' attention. The extensions are disposable to the moth, so even if a bat succeeds at biting them, the moth will simply break away and escape. It is a never-ending battle between predator and prey—one in which bats will surely adapt once again to refine their already stellar ability to echolocate. But let's see one ride a bike while doing that.

What do our pupils say about us?

Pupils dilate (expand) or contract as the light dims or brightens. But pupils also change size according to what the brain behind them is doing, whether that's recalling memories, analyzing a problem, or experiencing strong emotions. We may be unaware that our eyes are giving away so much while our brains are busy, but others who are aware can use that information to gauge their responses to us.

People have been deliberately sending messages with their eyes since at least the Renaissance. Back then, Italian women used eye drops derived from the deadly nightshade plant—which they called belladonna, or "beautiful woman"—to dilate their pupils, believing that it made them more attractive. It wasn't until hundreds of years later that anyone could figure out why dilated pupils would be so alluring. In the 1960s, a study showed that our pupils dilate when we're looking at something interesting or attractive. So a Renaissance man gazing into the eyes of a woman who had just used belladonna eye drops would see dilated pupils and unconsciously assume she was looking at something she found appealing: him!

What is it that the eyes love?

Eckhard Hess of the University of Chicago was responsible for those 1960s experiments, which were among the first to examine pupil dynamics. In Hess's studies, volunteers were shown images on a screen, and a camera photographed their pupils as they dilated or contracted in response to the changing pictures. The light levels were constant from one image to the next, ensuring that the changes in the volunteers' pupils were a response to mental activity rather than to light.

Hess was able to confirm the intuition of those Renaissance women: he found that men judge a woman's face to be more attractive when her pupils are dilated. Even when men were shown the same woman twice—the only difference being the diameter of her pupils—they preferred the image with the bigger pupils. Hess also confirmed that the phenomenon was more general than that. Pupils expanded when an individual saw anything interesting or attractive. But the same person's pupils contracted when he or she saw something unpleasant.

Did You Know . . . Another study by Eckhard Hess concluded that women who are attracted to "bad boys" (yes, they had a definition for that) responded most positively to males with dilated pupils. And another experiment, this one in the Netherlands, showed that people were more likely to give money to a virtual partner—and thus more likely to trust him or her in general—if that person's pupils were enlarged.

Our understanding of why pupils dilate has improved since Hess's experiments. We now know that pupils dilate in response to a range of mental activities, from recalling memories to making decisions while shopping or playing rock-paper-scissors. And it's not just our pupils that show our brains are at work. Blinking matters, too. Blinks signal the beginning of a mental process. After we blink, our pupils remain dilated as long as we're working on the problem. When we're finished, we blink again as our attention switches to something else and our pupils shrink.

The best data so far suggests that our pupils dilate the most when something is emotionally engaging. It doesn't matter whether that emotion is bad or good, just so long as it grabs our attention. In one experiment, participants filled out a survey asking if they were impulsive shoppers. They then watched a scene of people shopping. The researchers found that the people who identified as impulsive shoppers had the greatest pupil dilation—just viewing the activity of shopping was so emotionally exciting and stimulating for their brains that their pupils expanded.

That tight-knit connection between brain and pupils also happens when thinking is taking center stage. For example, while you try to solve the latest sudoku, you're constantly juggling numbers in your working memory. As your brain is managing those digits, your pupils dilate because of the mental effort. But if you were to stop concentrating and let your mind wander away from the puzzle, your pupils would return to normal.

We sometimes hear anecdotal reports of magicians being able tell what card you've picked out of a deck based on the size of your pupils, or clever shopkeepers knowing what you really want to buy by reading your pupils. But there's little research to support those stories. One experiment that came close to proving these claims used the game rock-paper-scissors to see whether people could predict their opponent's decisions by observing his or her pupils. Instead of playing the game person-to-person, the subjects watched video replays that showed a virtual opponent. Each subject was told that the video opponent's pupils would change when he or she chose an option (rock, paper, or scissors). Once the live players had been coached about what to look for, they beat the video players more than 60 percent of the time.

There was one problem with this experiment: because the subjects were watching a replay, they were seeing the video player's pupils dilate after his or her decision had been made. In a live game, players would have to act before that happened, so trying to use an opponent's eyes as a crystal ball wouldn't help most people win an ordinary game of rock-paper-scissors.

Poker, though? That's a different story.

Why does asparagus make my pee smell funny?

If you're one of those people who can identify a distinct aroma in your urine after you've eaten asparagus, then you're one of the lucky ones! Well, if not lucky, then you're at least among those who are able to smell it.

Over the years, many of the chemicals found in asparagus have been accused of causing the unusual odor, which can affect a person's urine when just a few spears are eaten. Although scientists had various culprits in mind, they did agree on one thing: something containing sulfur caused the asparagus smell.

But each time scientists thought they'd identified what those chemicals were, they were unable to explain how they could survive the long trip down the human digestive tract. The journey from fork to bladder was more than enough time for those compounds to be broken down and absorbed by the body, so post-asparagus urine should be scent free. If those chemicals are still present in a person's urine after dinner—and we know they can't have survived their trip intact—then it must be that some larger molecule survived and at the last minute changed into the odor-causing sulfur-based substance.

The best candidate for that large molecule is the appropriately named asparagusic acid. While asparagus is still in the ground, this acid serves as the plant's defense against nematode worms. There's more acid in the asparagus when it's young and still growing, which might explain why the pee odor is so much more pungent after you eat a young tip than an older, tougher stem. But the full details of the chemistry of the asparagus odor have yet to be worked out.

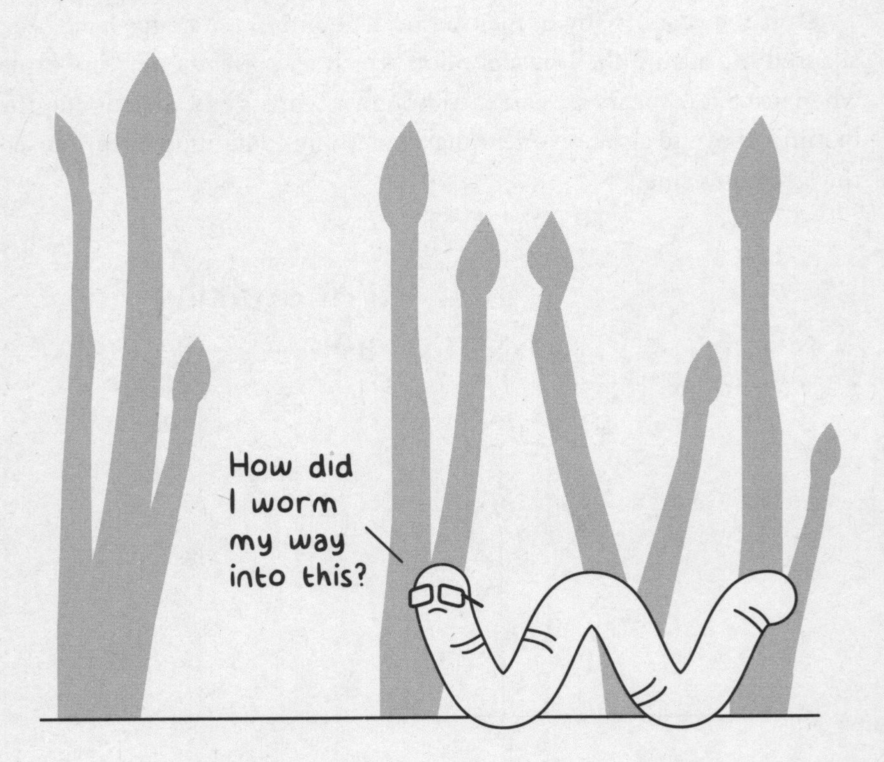

That's the accepted explanation for how asparagus gives urine that odd smell. But here's the catch: not everyone creates the scent. Experiments since the 1970s have produced conflicting results about how much of the population can produce this asparagus smell, with estimates ranging anywhere from 40 to 90 percent.

Did You Know . . . The prefix *thio-* in chemistry indicates the presence of sulfur in a chemical compound. Some of the chemicals in asparagus that are thought to cause the stink have long names, like *S-methyl-thioacrylate* and *S-methyl-3-thioproprionate*, but the *thio-* is a giveaway that sulfur is the real offender.

Similarly, not everyone has the ability to smell the stench. In one study, people were exposed to two samples of urine, each of which contained traces of asparagusic acid. The first set was original pee samples, while in the others, the asparagus scent had been diluted to an incredibly small 1 in 4,096 parts. A number of people in the study were unable to detect any stink in the undiluted urine, suggesting that they had no ability to smell the sulfur, while 10 percent were able to detect the asparagus smell even in the diluted sample, showing that they were hypersensitive to the odor.

A 2010 experiment added some detail. Volunteers ate a few stalks of asparagus and after a few hours submitted a urine sample. A couple of days later, the participants ate a piece of bread as a control substance and again deposited their urine. The researchers then recruited urine-sniffing volunteers, had them do their thing, and produced the most precise numbers yet: in roughly 8 percent of the asparagus samples, there was no sulfur smell at all. And when the smell was there, 6 percent of the volunteers couldn't detect it.

Science Fact! *Babe Ruth was at a posh dinner party when he was asked why he'd left the asparagus on his plate, an unusual move for a guy known to have a world-class appetite. Ruth replied, "It's just that asparagus makes my urine smell funny," or words to that effect. Ruth's line might have shocked the dinner table, but it was nothing compared to the words of author Marcel Proust, who, a few decades earlier, had declared that urinating after eating asparagus turned his "humble chamber pot into a bower of aromatic perfume."*

The biggest discovery from the experiment wasn't how many people were able to produce or smell the asparagus odor; it was how they were able to do so. The researchers found evidence of a gene that determines whether a person has the ability to smell asparagus in urine. As we've just seen, earlier experiments had shown that there's a wide range of sensitivity to the smell—to some people, the stench is overpowering, whereas others don't notice it at all. Knowing that, the scientists in the most recent experiment concluded that there couldn't be a single asparagus smell. Rather, there must be a variety of odors, each one as different as the person who produces it. That has yet to be tested, though—it's hard enough to persuade volunteers to sniff urine at all, let alone to judge its quality.

Do fingernails grow faster than toenails? If so, why?

Your fingernails, including the thumbnail, all grow at roughly the same rate, usually about 0.1 millimeters a day, or just under a millimeter a week. Toenails grow only half as fast, averaging a mere 0.05 millimeters a day, or less than half a millimeter a week. There's the first question answered—that was easy! But to understand exactly why fingernails grow faster than toenails, we need to examine how our nails grow in the first place.

Take a look at your thumbnail. At the farthest end, the nail loses contact with the finger and launches itself into space, providing an opportunity for dirt and infectious organisms to lodge themselves underneath. But as you move down the nail, you'll notice that the skin of your thumb surrounds and grows over it on both sides and at its base, forming a protective layer that prevents anything from getting into the soft tissue below.

Cast your eyes down a little farther, where the base of the nail meets your skin, and you'll see the white semicircle known as the lunula (the word means "little moon"). The lunula and other tissues around it, like

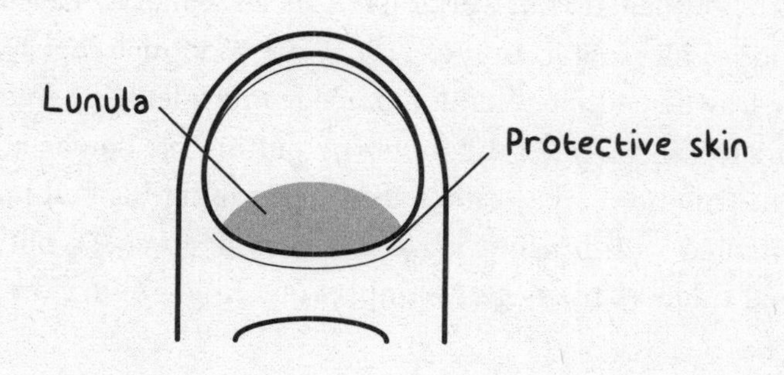

the fleshier-colored matrix, create the nail by continuously churning out cells; the fastest-growing skin cells anywhere in the body, in fact. Most of the nail is generated by the cells closest to the lunula. Those cells change, harden, and eventually die as they move toward the tip of the nail, leaving behind nothing but keratin, the main ingredient of the hard-shelled nail itself.

Did You Know . . . Have you ever noticed that your nails tear more easily across than down? Three layers of material make up the nail. The top and bottom layers are weak sheets of keratin that can tear in almost any direction. (They might not be strong, but they likely enhance the nail's ability to bend without breaking.) The middle layer is the strongest—six times tougher than the other two layers—and the keratin in it lies crossways, so it's four times easier to tear across than from the tip to the nail bed. And that's why your nails are easier to tear from side to side.

We all know that if you bash your thumb, the damage done will soon reveal itself as a mark on the nail, and that blemish gradually migrates out toward the tip of your thumb as the nail grows. Sometimes, if there's a tiny amount of bleeding, the mark left is what's called a splinter hemorrhage, a tiny straight line pointing toward the end of the thumb. These formations actually tell you something about the growth of your nail: the underside of the nail is striped with grooves that match similar channels on the surface of the tissue beneath the nail.

Given how nail growth works, it makes sense that damage to a nail can compromise its growth. In the early 1980s, a dermatologist in England, Rodney Dawber, suffered tendon damage to his left ring finger during a rugby match and treated the injury by putting his damaged finger in a splint. In true nerd style, rather than bemoaning his bad luck, Dawber experimented with his injury. He was curious to check out the idea of "terminal trauma," the negative impact of a finger injury on nail growth (even if that injury didn't damage the nail itself). For the three months he had the splint on his finger, he compared the growth rate of the nails on his two ring fingers, and he found that the growth on the injured finger slowed by about 25 percent. He also discovered (although this had been noted by others) that the nails on his dominant hand (his right) grew faster than those on his secondary hand. Meanwhile, his right and left toenails grew at about the same rate.

Dawber's experiment raises interesting possibilities, but it still doesn't definitively answer why some nails grow faster than others. There are just too many reasons why the growth of his injured finger's nail might have slowed. The so-called terminal trauma effect could have been the result of a reduction in circulating blood in that finger, or even simply the lack of stimulation that accompanies activity. To get a better idea of what factors affect fingernail growth, we need to travel back to before Dawber jammed his finger in a rugby scrum. Because, really, everything you need to know about the growth speed of fingernails can be traced back to one man: William Bennett Bean.

Bean kept a continuous record of the growth of his own nails from 1942, when he was thirty-two, to 1977, at age sixty-seven—thirty-five years! His work wasn't in vain, either. Through Bean, we learned that fingernails grow fastest when you're young and slow down over time. Warmth seems to stimulate the growth of nails, so if you live in a hotter climate, your nails will grow faster (although some recent studies have contradicted this). You'd think that would mean Canadians as a whole should see their nail growth speed up in summer and slow down in winter, but as Bean discovered, if you're working in a climate-controlled office that's warmer in winter and cooler in summer, the growth rate won't vary much. He also relayed information he garnered from other sources. For instance, he learned (à la Dawber) that immobilizing the arm or hand—such as by wearing a cast—will slow the rate of fingernail growth significantly, whereas being pregnant speeds growth. Bean even provided another angle on the Dawber injury theory. His own nail growth slowed considerably while he was ill with the mumps, showing that even indirect physical trauma can affect nail growth.

Bean presented all this data objectively—as is proper for a scientist, of course—but he admitted that watching his nail growth slow down over time hit him hard emotionally. (Dawber expressed exactly the same distress.) For Bean, it was the most obvious sign that he was getting old. While he could convince himself that he looked pretty good in the mirror, his nails were telling him something else.

Did You Know . . . Climate affects not only nail growth but also hair growth. In 1917, the dermatologist Felix Pinkus started recording the growth of a single hair on a mole on the back of his hand. The mole's hair would grow for a time before falling out. The hair regrew fourteen times over nine years. The hairs differed in lifetimes from 107 to 195 days—a huge disparity—and summer hairs lasted longer than winter hairs by an average of a couple of weeks. (Eventually, the mole changed and the hair disappeared for good.)

Now we come to the crux of the matter: Why do fingernails grow faster than toenails? It seems that they must continually be refreshed because they incur a lot of constant, if subtle, damage. The list of activities that might damage your nails is, at first glance, pretty slim. It's rare that we sink our nails into something or find ourselves clinging to a tree trunk by them. They're not talons or claws. But if you doubt that you're using your nails all the time, set an alarm and note what you're doing when it goes off. There's a good chance it involves using your nails. Even if you're typing, the nails reinforce the tip of the finger when it pushes down on a key.

There are even verbs in English that specifically describe actions you do with your fingernails: think of *scratch* and *pick*. If you're looking for more evidence, look no further than the fact that the nails on your dominant hand grow a little faster than those on your secondary hand. This tells us that healthy, active, well-nourished tissue is the basis for fast nail growth. Or consider that your middle finger is the longest and therefore the most exposed, and indeed its nail grows faster than the others.

By comparison, the toenails are pampered. For much of their active life they're in a protective shoe or slipper. Tapping or wiggling are the best bets for sustained, regular activity. Toenails rarely have to hold a pen or tie shoelaces, so if faster growth rate is a result of more frequent activity, there's no contest.

Another possible influence on the growth of nails is exposure to the sun. If that's true, toenails will indeed grow slower, given that they rarely glimpse the sun. You could argue that feet see less sun simply because of where they are on the body and the fact that they're so often shaded. Yet even in sunny parts of the world where toenails live a much brighter life, there isn't much evidence that they grow at the same rate as fingernails. And what data exists doesn't point to any difference between feet that are sun exposed and those that aren't.

A twist on this idea is that a nail's location on the body counts. Blood circulates more easily out to the arms than it does down to the feet, as returning blood to the heart from the legs is a much more difficult fight against gravity. The fact that it's more taxing for blood to circulate through the feet might also explain why toenails grow slower than fingernails. It's pretty much agreed that circulation is a factor, but the test of this theory would be to see whether such a diminished supply of fresh blood to the toes could be responsible for a nail growth rate that is less than half that in the fingers.

There is a tricky part to all this, too: toenails grow slower than fingernails, but they also grow thicker. The most common explanation for this is that toenails are constantly being punished by socks, shoes, kicking doors open and the like, putting the same sort of general stress on our toes that our fingers experience. But one question inevitably leads to another: If the constant activity and micro-injuries that are happening to your toes are similar to those happening to your fingers, why don't toenails grow quickly like their digital colleagues? Why thicker instead?

Science Fiction! *It's a myth that our nails continue to grow after we die. A scientific journal (although admittedly not a top-of-the-line one) once ran a study in which the author produced data showing that nails in cadavers continued growing for anywhere from eight to ten days! This has long been a popular horror-movie scenario, but no credible expert believes the results of that study. The apparent growth of nails after death is actually caused by the drying and retraction of the tissue around the nail, making them look longer than before. That puts a nail in that coffin, once and for all.*

Why do I get hiccups and how can I make them stop?

Let's take these two questions one at a time. In fact, the first question—why we get hiccups—is itself two questions: what happens in a hiccup, and why are our bodies are prone to them?

What happens in the body is a little complicated: it involves the brain and muscle groups in the chest, in particular the diaphragm. The diaphragm lies just above the stomach, and its movements enable breathing: when the diaphragm relaxes, it rises and air is forced out of the lungs; when it contracts, it moves downward and air is sucked into the lungs. When we hiccup, a part of the brain called the medulla signals the diaphragm to drop immediately. Normally, that would mean that you'd inhale a big gulp of air, but precisely thirty-five-thousandths of a second after that big gulp starts, the glottis, the space between the

vocal cords, closes abruptly, shutting down the incoming air. That's a hiccup, and the sound of it mirrors exactly what happens: a very, very brief inrush of air that's cut off almost immediately.

Hiccups are much more common in infants than in adults, and even more so in fetuses. In fact, fetuses just eight weeks old have been seen hiccupping in ultrasounds, for several minutes at a time, even though their diaphragms aren't yet fully formed. As we get older, we hiccup less and less.

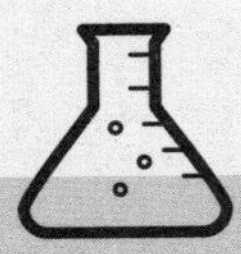

Science Fact! *Hiccupping seems to be confined to mammals. It has been seen in rats, cats, rabbits, horses, and dogs, but never in any reptile, bird, or amphibian.*

There are two main causes of hiccups. One is an immediate reaction to a stimulus. There's a range of things that can prompt a hiccup, including too much alcohol, too much food (especially if it's really spicy), bubbly beverages, and prolonged hysterical laughter. Most of these generate hiccups that might last for a few minutes or even an hour or two before they subside. But hiccups may also be triggered by a wide variety of medical conditions, including acid reflux, ulcers, a skull fracture, various infections, multiple sclerosis, and blood clots in the brain. In one bizarre case, hiccups were set off by a hair inside a person's ear canal tickling their eardrum. While it may seem odd that such a diverse set of conditions can cause hiccups, when you consider that the brain, nerves traveling between the brain and the diaphragm and many other muscles are all involved in a single hiccup, it makes more sense.

Long-lasting cases of hiccups are more common than you might think. One study that covered the years 1935 to 1963 found 220 cases of hiccups that lasted at least two days, with most continuing for more than two months. Men were nine times as likely as woman to be plagued with continuous hiccupping. Most persistent hiccuppers had periods of several days when the hiccups happened all the time, followed by days when there were none.

Did You Know . . . The world record holder for the longest continuous bout of hiccups is the late Charles Osborne, who started hiccupping in 1922 when he was twenty-eight years old and continued until 1990—sixty-eight years of hiccups! You might think that his hiccupping would have completely derailed his life, but he raised eight children and ran successful businesses. The sad irony is that after hiccupping for sixty-eight years, he lived only a few months after they stopped.

Why does the body have a mechanism for hiccups in the first place? Unlike vomiting, gagging, and coughing, hiccups are not very effective at expelling noxious substances from the body or clearing airways. The facts that hiccups are most common early in life and that only mammals hiccup prompted Dr. Daniel Howes at Queen's University in Kingston, Ontario, to suggest that the hiccup is crucial for mammalian infants. Howes argues that nursing infants can swallow a lot of air as they nurse, and that air can get in the way of incoming milk. By creating a bit of a vacuum in the chest, the hiccup draws that air up the esophagus, where it can then be burped out, leaving room for more milk.

Another provocative idea is that hiccupping is a hangover from a very distant time when tadpoles (or their ancestral equivalent) were adapting to breathing, either through gills or lungs. Switching from one to the other involves closing off the flow of air to the lungs, while opening up the flow of water through the gills, not unlike expanding the chest and closing the glottis. We have many neural programs that go back millions of years, why couldn't hiccupping be one of them?

Given the complexity of a hiccup, it's fitting that the list of hopeful remedies is long. Drug treatments are reserved for the most serious cases, but there are plenty of folk remedies, some of which are more scientific than others.

TRY THIS: Next time you have the hiccups, try swallowing ten times in a row, then holding your breath. This is my favorite, and the science behind this folk remedy makes sense: when you hold your breath, you raise the level of carbon dioxide in your blood. Exactly why carbon dioxide works isn't clear, but somehow it calms the diaphragm, neurologically speaking, making it less likely to trigger a bout of hiccups. The same thing happens if you breathe into a paper bag.

DON'T TRY THIS: My parents taught me that drinking a glass of water from the opposite lip of the glass cures the hiccups. In other words, tilt the glass away from you while you drink. The only problem: this usually leads only to more hiccupping and a lot of spilling.

The much-quoted *New England Journal of Medicine* published data suggesting that a spoonful of raw sugar would cure the hiccups. In one study, that worked nineteen out of twenty times, good enough to rank near the top of the methods in the study. In a 1999 article in the *Canadian Family Physician*, Ronald Goldstein claimed that having a hiccupper completely plug their ears and then drink a large glass of water through a straw would do it. Unfortunately, he quoted no statistics to support his theory.

In fact, a lot of studies of hiccup remedies are thin on numbers. In a research paper published in 2000, sex was suggested as the perfect remedy for the hiccups. But the study referenced only one case where this worked: a male who had been hiccupping continuously for four days.

Massaging the far back of the throat with a Q-tip for a minute is also said to have worked—but only once.

The most surprising remedy of all is surely digital rectal massage—meaning massage with the fingers. Even though there are three different reports published on digital rectal massage as a cure for hiccups, there are still only a handful of patients who've tried it. I suppose that's not surprising. Other than saying that there are a lot of nerves fanning out from the rectum and making connections elsewhere in the body, we are left guessing as to how this technique would alleviate the hiccups. And so far we are having trouble amassing a crew of volunteers for this research.

Why do my knuckles make that cracking noise?

We've all seen—or heard—it. You watch someone interlace their fingers and push their palms away from them. As the person stretches their fingers, they're rewarded with a sharp crack before they settle back down to business. To many people it's a painful sound, but the sound is actually a sign of something being created, not destroyed.

The cracking noise that a knuckle makes has to do with bubbles in the joints. From the 1940s until now, most experiments designed to figure this out have looked pretty much the same: only the imaging equipment has changed. In most experiments, a volunteer has a finger wrapped and tied to a cord that can be pulled to apply tension to the finger. Any finger will do; in the 1940s, the middle finger was used for most experiments, but today it's the forefinger. A force is gradually applied to the finger as images are recorded of the joint between the last hand bone and the first finger bone.

At some critical point, as the finger is stretched, the force being applied crosses a threshold and there's a sudden, explosive noise. From the early 1970s to 2015, it was believed that the cracking was the sound of a bubble (really a void) in the joint imploding. As the bones stretch apart, the argument claims, the bubble bursts into existence, then, just as quickly, collapses.

But Greg Kawchuk and his lab at the University of

Alberta have shown that it's the expansion of the bubble, not its collapse, that makes the sound. Kawchuk's images show that the birth of the bubble is just about immediate and that the suspected implosion of the bubble is more of a prolonged collapse. Their recordings of the crack show that at that exact moment of the sound, the space between the bones in the finger and the hand suddenly opens up, as much as doubling (0.04 to 0.08 inches, or 1 millimeter to 2 millimeters) in a tiny fraction of a second. That creates the bubble between the bones.

Before the finger is stretched, the finger bones are moving back and forth. They're in contact with each other, and any space between them is filled with fluid. When the bones are stretched, though, the separation is too sudden for more fluid to immediately flow into the space. So gases—especially water vapor and carbon dioxide—quickly enter the space from the surrounding areas instead, and a bubble forms instantaneously in the middle. High speed means big forces, and that's how the cracking sound is created.

Imagine trying to pull apart two sheets of wet glass stuck to each other; that sucking sound they make as they suddenly separate (which you can also make by squeezing the palms of your hands together and then suddenly separating them) is the same kind that's produced inside the joints of your fingers.

Any last doubts that this is a violent event were removed by a recent ultrasound study that revealed a bright flash of light, like fireworks, when a knuckle cracked.

Did You Know . . . It can take anywhere from fifteen to twenty minutes for the finger joint to ease back to its original spacing so that it can be cracked again.

All of this cracking and separating sounds painful, so it's natural to wonder if cracking your knuckles can damage your hands. For a long time, especially when it was believed that the sound was created by the collapse or implosion of the void (bubble), it was thought that cracking your knuckles could damage your hands. After all, the same mechanism erodes ships' propellers: bubbles form around the edges and implode, causing a shock wave. Over time that causes the metal of the propeller to fatigue. But despite the number of theories out there, the most damage that's ever been attributed to hand cracking is from a group of knuckle crackers studied in the 1990s who suffered from reduced grip strength and some swelling of the hands.

Did You Know . . . When Dr. Donald Unger was a child, his mother warned him that cracking his knuckles would give him arthritis. Determined to prove her wrong, Unger proceeded to crack the knuckles on his left hand twice a day for the next fifty years. He left his right hand alone during that time as a control.

At the end of his trial, Unger reported in an article to the journal *Arthritis and Rheumatism* that there were no signs of arthritis in either of his hands. He argued that this disproved his mother's claim, and it prompted him to wonder whether he should continue to trust her admonition that eating spinach is good for you. Critics later argued that his sample size (one) was too small to draw any conclusions.

So, the sound of your knuckles cracking is the sound of the birth of a giant (relatively speaking) bubble in the joint. That probably doesn't make it any less painful for anyone listening.

What is a hangover and how is one cured?

Humans first started drinking about eight thousand years ago, and the first hangover was probably the day after. Yet, surprisingly, there are still questions surrounding what causes hangovers and how they might be prevented.

People are going to drink, hangovers or not. (One estimate is that 75 percent of people who drink have had one.) One scientific study of hangover remedies in the *British Medical Journal* admitted that "no conclusive evidence shows that hangover effectively deters alcohol consumption."

While each person's hangover is uniquely unpleasant, there is a generally accepted (and very long!) list of symptoms, including: fatigue, headache, drowsiness, dry mouth, dizziness, nausea, sweating, anxiety, and a variety of mental effects, like the inability to concentrate and to remember.

Did You Know . . . As miserable as a hangover might be, it is not the same as alcohol withdrawal, even though many of the symptoms are common. People who drink to the point of intoxication just once can get a hangover. Withdrawal occurs when the habit of drinking large amounts of alcohol for a long time is suddenly stopped. Some of the most serious symptoms of withdrawal, like the DTs (delirium tremens, which includes hallucinations, sweating, shivering, high fevers, and even seizures) are not experienced during a hangover.

One striking fact: the hangover begins as the blood alcohol level declines, then peaks when alcohol disappears from the bloodstream and continues for as long as another twenty-four hours. Why is that? Two suspects are the metabolic products of alcohol and the other chemicals present in the drinks consumed.

Alcohol is processed into a chemical called acetaldehyde, which degrades first into acetate and finally into carbon dioxide and water. High levels of acetaldehyde are known to provoke facial flushing, increased heart rate, lower blood pressure, dry mouth, nausea, and headache, all of which fit nicely with hangovers. However, acetaldehyde is broken down quickly in the body, suggesting its role might not be that significant.

Experiments with rats suggest that acetate might be more important. The rats ingested pure ethanol and were tested for their pain threshold by probing around their eyes with short nylon filaments. Sure enough, they retreated from the filament at about the same time as the amount of alcohol in their systems was zero—in other words, just as the human hangover kicks in. They didn't react to acetaldehyde that way. The good news? Caffeine seemed to relieve the pain—interesting, since coffee was touted for its hangover relief more than a hundred years ago.

The other culprits in hangovers are substances called congeners, chemicals in various types of alcohol that are created in the fermentation or distillation process. There's a ranking system based on resulting hangovers: gin and vodka have the fewest of these additional chemicals, while red wine, bourbon, and especially brandy have the most.

Of all human behaviors, drinking might be the one that's most resistant to that ultimate of study designs: the randomized double-blind trial, in which participants receive a placebo or alcohol, but neither they nor the experimenters know which they're getting. We've all heard stories of people at a party who are fooled into thinking they're drinking and who then act drunk, but in a lab setting? Do participants consuming a placebo not know that's what they're drinking? Can the experimenters not tell? The difficulty of establishing what exactly provokes symptoms stands in the way of coming up with a cure.

While a Google search for "hangover cures" or "hangover remedies" generates hundreds of thousands of hits, a quick glance reveals that almost none have been scientifically evaluated. No surprise.

One scientific review contrasted a short list of hangover remedies touted on the Internet (including aspirin, fresh air, honey, cabbage, and at least three versions of drinking more alcohol) with a much smaller set of actual studies and concluded that even the actual published-in-the-scientific-literature studies didn't amount to much. Those that hinted at some effective hangover treatment, like prickly pear, artichoke, or dry yeast, either didn't include enough subjects or had some other constraint that required the study be repeated. The authors concluded there was no evidence for any effective intervention.

So I'll end with another quote from the same article in the *British Medical Journal*: "The most effective way to avoid the symptoms of alcohol induced hangover is to practice abstinence or moderation."

Could humans ever hibernate?

When we think of hibernating, we think of bears fattening up, finding a den, and sleeping all winter. But there are many ways to hibernate, and there may be some compelling reasons why humans would want to do it. The most obvious application of hibernation is for keeping patients in a slowed-down state for transplants, for prolonged surgery and recovery, or even for space travel.

A typical bear hibernates for anywhere from five to seven months. During that time, its core body temperature drops 8 degrees Fahrenheit (5 degrees Celsius) and stays there for weeks at a time. The body shuts down to about two-thirds of its normal activity and the heart slows from forty beats per minute to ten to fifteen. Even though they're hibernating, bears still burn a few thousand calories a day.

They can lose up to a quarter of their body weight during hibernation, almost all fat, and yet they suffer no bone loss or muscle wasting (whereas bedridden humans certainly do). Bears don't pee or poo when

hibernating. Their kidneys nearly stop but don't fail, and they don't accumulate deadly levels of chemicals.

Did You Know . . . When bears emerge from their hibernation in the spring, they don't eat or drink much for the first couple of weeks, even though they've taken in nothing for months. That changes quickly, though. By midsummer a good-sized bear consumes 5,000 to 8,000 calories a day, and grizzlies, just before hibernation, take in 1,000 calories an hour, twenty hours a day.

Bears grab our attention, but when it comes to hibernation they're not nearly as impressive as some smaller animals. Arctic ground squirrels in Alaska hibernate for anywhere from eight to ten months. As they hibernate, the temperature in some parts of their bodies can drop as low as a degree or two below freezing. That's impressive, but some animals can survive being totally frozen. For example, if wood frogs are dug up in the winter, they are exactly like ice cubes: if you dropped one on the floor, it would shatter. Ground squirrels don't turn into chunks of ice like this, though. Instead they're just supercooled: the water in their bodies stays liquid instead of freezing. This is a precarious state: you can put a container of water in the freezer and sometimes it will remain liquid until you jostle it or even just tap the side with your finger. Any disturbance like that triggers the instantaneous formation of ice. Rats and hamsters can be supercooled, but if they're left in that state for more than an hour, the water in their bodies will start to turn to ice and crystallize and they'll die. Somehow, Arctic ground squirrels can maintain that awkward state for days. It's not clear yet how they do this; they might be able to clear their bodies of any particles around which ice crystals could form, but no one knows for sure.

These squirrels also differ from bears in that they wake up every few weeks to raise their body temperature, move around a bit, pee and poo, then return to hibernation. These brief awakenings must be necessary for the animal's survival, because it costs a hibernating animal a lot of energy to return to normal life and then slow down again. Maybe these awakenings have something to do with the extremely low temperatures they reach.

Looking at bears and squirrels, you might assume that hibernation is all about escaping the cold. But that's not always true. The fat-tailed lemur—an animal living in Madagascar, one that is more closely related to us than either bears or squirrels—hibernates for about seven months of the year even though the temperature in its tree den can range from 50 to 86 degrees Fahrenheit (10 to 30 degrees Celsius) during that time. The seven months that the lemur is holed up are the driest months, when food is scarce, and apparently it's just not worth it for the lemur to patrol the forests searching for it. Instead, it prepares for hibernation by practically doubling its weight in the plentiful months.

The fact that animals' bodies remain healthy through the dramatic changes of hibernation is amazing enough, but if we want to know if humans could hibernate, we have to focus on the brain: it's the most energy-hungry organ, and a hibernating animal does everything it can to reduce energy consumption.

Science Fiction! *Hibernation is not sleep. In fact, hibernation might actually deprive animals of the kind of sleep they need most. When Arctic ground squirrels break out of hibernation every few weeks, they use a big chunk of their non-hibernating time to sleep. It seems that sleep is something a hibernating animal can't afford to do, because sleeping burns too much energy.*

A slowing metabolism has dramatic effects on brain cells. As a hibernating animal's body slows, the branching that allows each brain cell to communicate with thousands of others shrinks and retreats, and the skeletal system that supports them starts to collapse. Some of the changes seen in hibernating brains actually mimic destructive changes seen in the brains of Alzheimer's patients. But when the animal wakes, its brain kicks off an amazing bout of activity, reestablishing connections and restoring skeletal frameworks. It's not completely clear whether these changes affect the animal's memory or not.

Did You Know . . . A Japanese hiker named Mitsutaka Uchikoshi survived twenty-four days on the side of a mountain without food and water after he passed out, having suffered a broken pelvis as he was hiking. When he was found, his body temperature was 71.6 degrees Fahrenheit (22 degrees Celsius). He should have been dead, but he survived virtually undamaged. Although this wasn't the same mechanism that a bear or an Arctic ground squirrel goes through, doctors were quick to say that his survival was due to something like "hibernation."

We have not evolved to hibernate, so some features of hibernating animals, like maintaining a heartbeat at 33.8 degrees Fahrenheit (1 degree Celsius) are impossible for us. The human heart fails when body temperature drops below 68 degrees Fahrenheit (20 degrees Celsius). Like the lemur, we'd have to arrange hibernation at a reasonable temperature. And we would have to figure out ways of ensuring that we would wake up from hibernation every few weeks—as the Arctic ground squirrel does—so that we could eat, drink, pee, or poo (or all of these). That might also help with the issue of sleep deprivation. If our brain cells started to lose connections and our skeletons started to fall apart, we'd need to be absolutely sure that everything would recover when we woke.

Even so, human hibernation might not be impossible. There could be a way of shortcutting millions of years of evolution to provide a safe, human-only form of hibernation. But the subtlety and sophistication of the mechanics of it suggests it's a long, long way off.

Do we all have Neanderthal in us?

I love the Neanderthals, and apparently if I had lived several millennia ago, I might have meant that literally! Neanderthals are the other human species, the "cavemen" who lived in Europe from about 350,000 to 40,000 years ago. Neanderthals have typically been portrayed as primitive, clumsy, heavily built, knuckle-dragging, unintelligent brutes who had scraped out an existence in Ice Age Europe before dying out when confronted by the graceful, more intelligent, craftier modern humans.

It's true that Neanderthals were, on average, stockier than modern humans. They were muscular and powerful, built for rapid movement from side to side. Compared to modern humans, Neanderthals looked quite different. They had relatively large noses—in fact, the nose and cheekbones were slightly pulled forward compared with ours. Their larger sinuses might have been designed to warm inhaled air, a necessary feature in a cold-climate hominid. Their brains were bigger and heavier, and they had a different shape, with a prominent bump at the back. It's

been speculated that this was an enlarged cerebellum, the part of the brain that coordinates movement, among other things. That would make sense, given that much of the Neanderthal brain space was devoted to vision and muscular movement, with less allotted to the frontal lobes, where our ability to plan and make decisions resides.

In a perfect example of how a single discovery can set an inaccurate tone for nearly a century, a Neanderthal skeleton excavated in France in 1908 was described as being hunched over, more animal than human. Years later, closer study of the remains revealed that the excavated individual had been crippled by arthritis in his spine. It was the arthritis, then, and not some characteristic feature of his species, that accounted for his inability to stand erect.

The new information didn't help, as the Neanderthal's unimpressive résumé had been set in stone. The old saying that the victors write the history books has never been truer than for the Neanderthals. At first, the fact that we were still here and they weren't seemed to tell the whole story. Shortly after they encountered modern humans (us), the Neanderthals slowly withdrew to a few final tiny outposts in southern Spain and Portugal before disappearing for good, and that the Neanderthals didn't last was thought to prove their inferiority.

Did You Know . . . Neanderthals have a sister group, called the Denisovans, a mysterious, newly discovered group that lived around the same time as the Neanderthals—tens to hundreds of thousands of years ago—and apparently spread their genes around, too.

In the last twenty-five to thirty years, though, the Neanderthals have undergone a radical image rehabilitation, to the point that what made them different from us seems much less obvious than what made them the same.

Culturally, Neanderthals have always seemed a bit backward compared to modern people. Our ancestors—the modern people of the Neanderthals' era—are responsible for most of the thirty-thousand-year-old cave paintings in southern France and Spain, but they don't have a monopoly on art. There is an example in Gibraltar called Gorham's Cave, where researchers have found tic-tac-toe-like scratchings on a wall—abstract art, some call it—that date to when the cave was occupied by Neanderthals. (Gorham's Cave is apparently one of the last places they lived.) And it's not just the cave art that hints at culture: there's solid evidence that Neanderthals buried their dead, used fire, and decorated their bodies with teeth; claws; feathers; and red, black, and yellow pigments. They also hafted blades onto spears and were capable of crossing open water (presumably voluntarily).

The Neanderthals were hunter-gatherers, and so they had to go where the food was. Evidence of the Neanderthal diet is gleaned from chemical analysis of micro-traces of food left behind in the plaque on fossilized teeth. Neanderthals consumed a wide-ranging diet that was dominated by meat—there are butchered animal bones aplenty wherever their remains have been found—but that also included vegetables. Chemical analysis of a famous Neanderthal skeleton from Saint-Césaire, France, suggested that he and his cohort ate fewer reindeer and more woolly rhinos or mammoths than did their main competitors, hyenas. The rest of their potential prey—deer and horses—they split fifty-fifty with the hyenas. There are also hints of Neanderthals consuming some plants for medicinal purposes.

Canadian scientist Valerius Geist came up with the most insightful and spectacularly visual account of how Neanderthals managed to hunt and kill the biggest mammals of their time: rhinos and mammoths. The Neanderthals' spear points were coarse and heavy, unlike the finely shaped versions preferred by our ancestors. The Neanderthals' craftsmanship has often been seen as a deficiency attributed to their supposedly primitive skills, but Geist argues that those clumsy spearheads were perfect for the job. He claims that Neanderthals worked at close quarters when hunting, and that only two men were needed. One Neanderthal would approach the animal from the front (carefully) and irritate it. At the same time, hunter no. 2 would close in from the side and grab the mammoth by the fur. (Mammoths and rhinos of the time had long hairy coats, and Neanderthals had exceptionally strong hands with long fingers and strong, broad fingertips.) The animal, disturbed by the thing clinging to its side, would wheel and buck, trying to shed the nuisance. Hunter no. 1 would take advantage of the distraction to drive his spear into the animal. The lighter, slimmer spears preferred by our ancestors were good only for throwing; in close quarters, they likely would have broken off on a bone. Two Neanderthals working right next to the animal, then, could do the work of five modern humans throwing from a distance.

Geist compares the Neanderthal hunting style to the way that rodeo clowns distract bulls. In fact, the pattern of bone breaks seen in Neanderthal skeletons mirrors closely the typical injuries suffered by rodeo clowns today. A study in 1995 compared bone breaks among modern humans living in New York and London, Native Americans from hundreds to thousands of years ago, Neanderthals, and rodeo clowns. The Neanderthals and the clowns stood out from the rest in that they both had a much higher percentage of head and neck fractures and fewer lower-limb breaks. It's not proof, but it's pretty suggestive that Geist—who got his idea of close-encounter hunting while watching bull riders and clowns at the Calgary Stampede in 1973—might be right.

Whether or not Neanderthals used language is a long-running controversy. Earlier theories argued that they died out because they didn't have the brain machinery for language, spoken or signed, and therefore couldn't compete with the more highly developed modern humans. One key piece of that argument was that Neanderthals' throats didn't appear to have a bone, the hyoid, that plays a crucial role in articulating speech. But—in yet another example of how the picture suggested by advancing science is always changing—a sixty-thousand-year-old Neanderthal skeleton found in Israel apparently has a hyoid identical to ours. That doesn't mean that the individual, or its species, could speak, but at least there's one less barrier to that possibility. Unfortunately, speech leaves no fossil traces.

To understand whether modern humans have any Neanderthal in them, we must turn to genetics. The technology that allows us to reconstruct large portions of the Neanderthal genome is staggering and brilliant. The first sets of Neanderthal genes were extracted from a fraction of a gram of bone! Scientists have even been able to use genetic analysis to show that some Neanderthals would have been fair-skinned redheads.

The bottom line is that, yes, in those of European descent, up to 5 percent of the genome can be Neanderthal genes. Five percent of the modern European genome sounds like a lot, but remember that a majority of those Neanderthal genes probably code for things we're never aware of, like some tier-2 protein in the kidney. It's tempting to think of the Neanderthal contribution as the "gene for the big nose" or "the gene for muscular legs," but so far there is no evidence that we have Neanderthal genes that tweak physical differences.

That's not to say that Neanderthal genes aren't useful. There's a cluster of them on our chromosome 3, and some of those play a role in adapting to the ultraviolet light in sunlight. Those particular genes are even more common in East Asians' genomes. Somehow, modern people acquired a set of UV-light-adapted genes outside of sunny Africa that were more effective than the ones they'd had upon leaving it. A second, vital set of imported Neanderthal genes is found in the HLA system, a group of genes that gives the immune system the ability to identify and defend against invading bacteria and viruses. The protein molecules produced by HLA proteins can confer resistance against infections, including the Epstein-Barr virus, which causes infectious mononucleosis and is associated with a cancer called Burkitt's lymphoma.

Neanderthal genes are even more common in East Asians, but they're effectively absent in Africans. This has led to the conjecture that the interspecies hanky-panky happened mostly in the Middle East, especially the Levant, the area around the east coast of the Mediterranean. We'd all like to know what those gene-swapping encounters were like. It's well known that the human brain is very good at evaluating others as "us" or "them," and that reaction had to be heightened when modern humans met the Neanderthals. But did the reaction intensify the exotic or tune it down? Was sex consensual? So far, the science doesn't allow us to say definitively whether the matings were between Neanderthal males and *Homo sapiens* (us) females or vice versa, but the former seems most likely. That's partly because in this case, males would be more likely to be infertile, leaving no trace of themselves, and also because genes that would be unique to female Neanderthals—mitochondrial genes—are absent in modern populations.

And how often might these interactions have happened? We don't know. I've seen numbers ranging from one mating per several dozen encounters—a mere handful per year—to totals of several hundred or even a few thousand. Not an orgy, but apparently often enough to leave behind evidence that has lasted more than forty thousand years.

Whatever the exact number is, one thing is certain: the notion of a Neanderthal dragging any woman back to his cave by the hair is nothing but myth. What sort of evidence would one need to be able to believe this? A stretched-out skeletal hand with long hairs in its grasp? Consistent sets of micro-fractures on female Neanderthal skulls? There isn't any such evidence.

The earliest reference to this idea is from the nineteenth century. One Andrew Lang, a prolific and wide-ranging writer and commentator, wrote about nomadic life in Europe just as the ice sheets were retreating. He had this to say about that period: "In the big cave lived several little families, all named by the names of their mothers. These ladies had been knocked on the head and dragged home, according to the marriage customs of the period, from places as distant as the modern Marseilles and Genoa."

Out of the mind of an imaginative man directly into the public awareness. Lang didn't specify that this description was about Neanderthals—though they had been discovered by then—but once the Neanderthals became popular, it was easy to fit them into that scenario. It took many years to dispel the notion and prove that it's just the stuff of cartoons.

Did You Know . . . Neanderthals have traveled across time and space in pop culture. English author William Golding is best known for his first novel, *Lord of the Flies*. His second book, *The Inheritors*, described an encounter with and subsequent elimination of a small group of Neanderthals by a band of modern people. His Neanderthals were a curious breed, guided by pictures in their heads, incapable of much language and perplexed by the bows and arrows used against them. Of course, Golding was speculating (this was 1955), but although he was likely wrong about the archery, there is evidence that Neanderthals had larger eye sockets and perhaps were a more visual and less linguistic people than we are.

Years later, science fiction author Robert J. Sawyer wrote the Neanderthal Parallax, a trilogy in which the Neanderthals survive in a parallel universe. Eventually, one individual somehow crosses between these universes to meet modern humans at the Sudbury Neutrino observatory, in Ontario. He would be the only Neanderthal ever to know anything about quantum physics—as far as we know.

Even after 160 years of study, no one is really sure why the Neanderthals died out after managing to exist across Europe and the Middle East for hundreds of thousands of years. The traditional view—now largely abandoned—was that modern humans killed them off (admittedly a very human thing to do), and it's true that the timing of our ancestors' arrival in Europe is suspicious: the latest dating techniques suggest that modern people arrived in Europe about forty-five thousand years ago, and the Neanderthals, although already well established, were pretty much finished five thousand years later. True, five millennia is a long time over which to extinguish a species, but both populations were sparsely distributed, so contact might have been rare. The problem with this theory is that there's no fossil evidence of pitched battles between the two closely related hominids, and although modern people supposedly invented both the throwing spear and the bow and arrow, you'd think that hand-to-hand battles would favor the more physical Neanderthals. Some have conjectured that it was tropical diseases we brought from Africa that did the deed.

But the preferred explanation these days points to a combination of environmental changes. As the glaciers continued to recede and their favorite prey dwindled, the Neanderthals found themselves unable to cope with those changes and also compete with us moderns. But although the Neanderthals as a species disappeared, the fact that a portion of their genome clings to ours shows that their rodeo-clown legacy is still alive and well.

History Mystery

Did Newton really get hit on the head by an apple, inspiring his thoughts of gravity?

This is the best-known science story ever: Newton, sitting in the garden, musing under an apple tree. An apple falls from the tree and inspires him to invent a whole new way of thinking about gravity.

Did it really happen? Unfortunately, Newton himself was mostly mum on the subject. He never said anything about this moment (which supposedly happened when he was in his early twenties) until he was nearing death, at the age of eighty-four. At that time, sitting in a different garden with his old friend William Stukeley, he explained how he'd begun to question gravity based on the way an apple falls: Why always straight down? If it were just Stukeley's claim to have heard Newton say this, it might have been

forgotten, but three other writers claimed Newton told them the same story. And as with all stories, there are inconsistencies—for instance, one person mentions the garden but not the apple.

Science Fiction! *While Newton may very well have been sitting under an apple tree when a piece of fruit fell, inspiring thoughts about gravity, he was never hit on the head by that apple. That part seems to be totally fictitious. But it does make the story more colorful!*

The notion of gravity strikes us now as common sense, but in Newton's time, it wasn't. In the 1660s, Newton had returned home from Cambridge University because of the threat of the Black Death. He was thinking hard about gravity, a force that to most people was mysterious. There were only vague and somewhat mystical explanations about why things moved the way they did, such as the idea that invisible strings were attached to everything, and heavier things had more strings. Or the theory that while some things obeyed gravity and fell to earth, others, such as smoke, were subject to "levity" and rose instead of falling.

The grand ideas of attraction and movement were on Newton's mind, and so it isn't unreasonable to think that the sight of an apple falling to the ground might have caught his eye. But that's not the amazing part. What's amazing is what he did with the idea. He told Stukeley that it occurred to him that the earth draws the apple toward it, but the apple must also draw the earth. He also linked the fall of the apple to the movement of the moon in the sky, concluding that they were both examples of the same force. The moon was, in effect, falling toward the earth, but it never hit because its speed carried it around the earth in an endless lunar orbit. This thinking, rendered in complex math and extended to the visible universe, was the basis for Newton's mighty *Principia*, in which he laid out his famous laws of motion in 1687.

Just to put this in an even larger and more impressive context: Newton was either twenty-three or twenty-four when that apple fell. He was thinking about gravity but also busy inventing calculus and figuring out the nature of light (determining that light was composed of the colors of the rainbow). Absolutely mind-boggling.

Not all Newton scholars love the story of the apple. One has written that it is "vulgar" to think that the mere fall of an apple inspired the mighty laws of gravitation. But the apple wasn't all there was to it: it was another twenty years before Newton published all the complex ideas that resulted from his thoughts about gravity. Also, what's the problem with a commonplace thing being inspirational? That inspiration still requires a great mind. In this case, one of the greatest ever.

Science Fact! *Newton's apple tree, at Woolsthorpe Manor in England, has a colorful history. It was the only apple tree in the garden at the time that Newton rested there, and a century and a half after its most famous apple fell, it was blown down in a storm. It managed to survive long enough to procreate, though. Cuttings from the tree and its descendants have been taken many, many times and grafted onto apple trees all over the world. In Canada, there are several descendants of Newton's tree, the first of which was planted at York University in Toronto.*

There is still a mystery, though, because genetic analyses reveal that two trees have given rise to all of Newton's trees around the world. That means that some proud owners of this vestige-of-a-great-moment-in-science may have nothing more than a regular apple tree. The trees do produce apples, though—a variety of cooking apple called Flower of Kent, said to be good for cooking, not eating, because it is mealy and flavorless.

Part 2
The Human Brain

Why are yawns contagious?

About 50 percent of us are susceptible to contagious yawns: that means we feel the irrepressible urge to yawn as soon as we're aware that someone else is yawning. I say "aware" because you don't have to actually see a yawn in order for it to be transmissible. A contagious yawn can be triggered by hearing a yawn, overhearing a person talking about a yawn, or even just reading about one. The yawner could be in a video you're watching, and the video could be sideways or upside down. If you're one of these people who falls prey to the contagious yawn, you know that it is never really your choice to yawn; you can't stop it. But why do we share this behavior? What's the point?

Robert Provine at the University of Maryland is one of the world's most prominent yawning researchers, and he has established the basics of

the process. On average, an ordinary yawn lasts six seconds and involves a huge inhalation followed by an exhalation, stretching the mouth open to its limits and squinting the eyes. There's a repeatable series of events in a yawn—eyes close, mouth opens, air moves in and then out, yawner relaxes. Once you start the sequence, it's hard to stop it, and if you do, you feel unfulfilled. You can yawn while pinching your nose closed, as odd as it feels. But try it with your teeth clenched and you'll find a yawn is very difficult, if not impossible, to complete. That's curious, because if the purpose of a yawn is to move a lot of air in and out, you can do that perfectly well with your teeth clenched. Conversely, you can have a huge intake and output of air and still have a failed yawn. You need the whole package of facial actions to propel your yawn to completion.

Provine has tried to identify exactly what parts of the yawning face trigger the contagion of a yawn. Surprisingly, the mouth isn't that feature: yawners whose mouths are obscured nonetheless prompt yawns in those watching them. The flip side is that a yawning mouth on its own isn't easily recognized as yawning—it could just as well be yelling. Also, if you cover the mouth during a yawn and just watch the other parts of the face, then a yawn looks a lot like an orgasm. The two have similar dynamics—the buildup to a climactic moment, followed by the return to a baseline—suggesting they might be sharing fundamentally similar low-level brain mechanisms, ones that have been around for millions of years. No one has yet addressed whether you can yawn and have an orgasm at the same time. Maybe you can't because there's only one set of facial muscles available for both?

More than just tracking the mechanics of a yawn, though, Provine has disproved the common theory that yawning is a response to inadequate oxygen or too much carbon dioxide in your blood. He's shown beyond doubt that neither reducing oxygen nor increasing carbon dioxide increases the frequency of yawning. According to Provine, the epidemic of yawns triggered by a stuffy lecture hall in the late afternoon has less to do with the air in the room and more to do with the lecture itself. Provine tested this idea. He found that students watching static on a screen yawn more often than their counterparts watching videos. It was boredom and fatigue, then, that led to increased yawning.

That makes sense in the late-afternoon lecture hall. But it's hard to see why boredom, which comes in many forms, would instigate a behavioral response as universal as a yawn. Restlessness, yes. Going to the fridge, maybe. But opening your mouth wide and closing it? There's no obvious connection.

Did You Know . . . Scientists in China and at the University of Michigan are developing software that can analyze images from a dashboard camera to detect driver fatigue. Yawning is a reliable clue to fatigue, but capturing a moving driver's face in variable light conditions isn't easy. The system is designed to detect the indicators of yawning while ignoring the mouth. The setup so far includes a face detector, a nose detector, a nose tracker (to follow the nose's movements), and a yawn detector. Scientists working on a similar detector at the University of Strathclyde are also concentrating on the rest of the yawning driver's face rather than the mouth, on the sensible grounds that people do often cover their mouths when they yawn.

Another theory is that yawns cool the brain. This could be useful in warm weather—and there is evidence that people yawn more often in warm temperatures than cold—or when the brain literally heats up if you've been thinking hard. Rats with mini-thermometers in their brains exhibit a quick rise in temperature immediately before a yawn, although this effect has been questioned because the rats are yawning and stretching simultaneously, so it's hard to know which causes the spike in temperature. It's also not clear that the cooling effect lasts for any appreciable time.

In the end, it might turn out that contagious yawns have both physiological and psychological causes. Some psychological explanations trace yawn contagiousness all the way back to early hominids. According to those theories, yawning was supposed to be a signal passed around the group, to heighten awareness, to start moving together, or to bed down for the night. There's evidence that passing yawns around a group establishes empathy. That's a tricky proposition to test directly, but there are some clues. Children don't generally participate in contagious yawning until they're four or five years old, when they acquire something called theory of mind, which is essentially the ability to understand that others are having thoughts and emotions of their own. Children have been yawning spontaneously since they were in the womb, though, suggesting there are two different mechanisms in play—one for a standard yawn and another for a contagious one.

Acquiring theory of mind is a crucial development for social beings like us. Some children with autism spectrum disorder are slower to (or never) achieve theory of mind, and that affects the development of their social lives. They are not susceptible to contagious yawning. Neither are people who, while not full-on psychopaths, have some psychopathic traits, like selfishness, manipulation, impulsiveness, callousness, dominance, or, above all, lack of empathy. The more psychopathic traits an individual displays, the less likely he or she is to yawn in response to someone else's yawning. One 2016 study found that females are more susceptible to contagious yawns, a result that was immediately recruited to support the "empathy" idea.

Other explanations for why yawns are contagious take a wildly different approach, looking at how yawns occur in other species. Chimpanzees frequently respond to a yawn with one of their own, but they do so only in response to yawns from other chimps, not humans. High-status baboons yawn a lot, and their yawns have the happy coincidence of exposing their giant, threatening canine teeth. Baboons on the lower social rungs are usually wise enough not to yawn back, as contagiousness in those situations might be fatal. It was once suggested that this threat avoidance was the origin of covering your mouth when you yawn. The problem is that, even if you did cover your mouth, the aggressor might still know you were yawning just by looking at the rest of your facial expression.

A 2008 study stirred a lot of excitement by claiming that dogs were prompted to yawn when they saw other dogs, or people, yawning, and that this reaction occurred at a higher rate than it did in humans. This was a major coup for the empathy idea, given that dogs have made it to where they are today by exploiting empathy. Unfortunately, other studies have failed to confirm the original findings, so we're left with a possibility that dogs respond to yawns, too, but really no definitive evidence to back up the claim.

Did You Know . . . Budgies yawn, but they're the wild card of contagious yawning. When budgies are allowed to perch across from each other, separated only by glass, their yawns cluster together in time. The response yawns are not rapid-fire by any means: some of them take minutes to appear. But that's still significantly different from when the budgies can't see each other; in that case, any yawns are not connected time wise at all. Birds are a lot more intelligent than we have given them credit for, but an empathetic budgie really stretches the list of animals known to experience contagious yawning: humans, chimps, dogs, and high-frequency-yawning Sprague Dawley rats.

There has been some mapping of what goes on in the human brain during a yawn, and one revealing discovery was the absence of activity of mirror neurons. Mirror neurons are brain cells dedicated to monitoring the behavior of others and imitating it. That they are not involved in yawning suggests that the signals kicking off the yawn are coming from somewhere deeper in the brain. This would mean, first, that it's more or less automatic and, second, that it likely goes far back into our ancestry. It is a deep-seated behavior.

Stifling a contagious yawn is different—you have to think to do it.

Cells in the prefrontal cortex, which are involved in thoughts and actions involving empathy, are active during contagious yawning. The prefrontal cortex is silent during spontaneous yawns, though, suggesting a different neural circuitry is involved in the two kinds of yawning. How remarkable it would be if it turned out that over the course of our evolution we co-opted an established physiological reflex for shaking ourselves out of fatigue and turned it into an empathy signal?

Does subliminal advertising work?

The quick answer is this: under very tightly controlled conditions, subliminal advertising can work. The catch is that the circumstances have to be so tightly controlled that it likely isn't worth taking the time to create them. Without the ideal conditions in place, subliminal advertising just isn't all that effective.

A subliminal message is one that is presented on a screen so quickly that you're not consciously aware of it. These messages can be measured in thousandths of a second. Although these sorts of messages are too short to be consciously recognized, they can still have an effect on your brain, because most of what's going on in there is unconscious. Your consciousness—your awareness of what's going on in your own head—is actually only a tiny fraction of the second-by-second activity in your brain.

Both lab experiments and analyses of consumer behavior suggest that more often than not, we're acting on behalf of our unconscious mind. In one experiment, a group of students was exposed to a long list of words, some of which suggested old age, such as Florida, bingo, and forgetful. On a similar list for a second group, the "old age" words were swapped out for ones that don't suggest age, such as California, awkward, and chess. The study found that the students who had been given the "old age" terms walked more

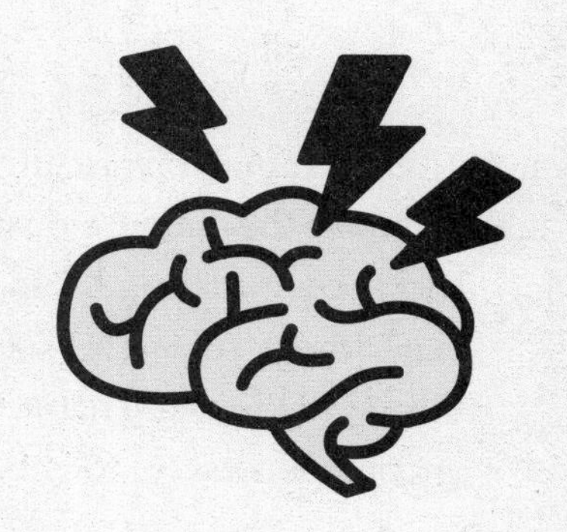

slowly when they left the room, but they denied being consciously aware of their slow gait. To the researchers, that seemed to prove that the students' unconscious minds had picked up on the cues in the terms and were guiding their actions accordingly. But a swirl of controversy—still unresolved—broke out when subsequent researchers said they'd been unable to reproduce those results.

A less contested experiment in the Netherlands showed that purchases of major items—things like cars—don't necessarily require a lot of thinking or analysis. In the test, researchers gave volunteers a set of characteristics of different cars, and then they either distracted the buyers with mental exercises or allowed them to quietly contemplate the purchase. Those who had been distracted—meaning they weren't able to think about the purchase—made the most logical choices, while the people who thought about the purchase long and hard selected lower-quality vehicles. The study showed that when it comes to buying items with fewer features to consider, such as oven mitts or toothbrushes, it's fine to use your conscious mind to decide what choice is best. But when the decision is a complicated one, the unconscious mind might actually be more reliable.

Did You Know . . . The same research group that ran the experiment about car purchases also ran a study about furniture. They found that people who mulled over decisions in Ikea—rather than simply going in and picking the first item they liked—were less happy with their choices in the long run.

Subliminal advertising messages seem like a logical step from there: persuade the unconscious mind to buy the advertised article. The research shows such messages can work, but only under tight constraints. In one experiment, researchers showed a group of volunteers a subliminal ad that displayed the brand name of a soft drink. The scientists hoped the ad might encourage viewers to choose the advertised product over a competitor's.

What they found, though, was that the messages worked only if the volunteers were thirsty. Not only that, but if the advertised brand was already the person's preferred drink, the subliminal message didn't increase his or her desire for it. The experiment did find that if people were thirsty and they didn't have a strong preference for any particular drink to begin with, they would tend to choose the drink that was subliminally advertised. But the conditions in the lab were carefully controlled—nothing like the real experience of either TV watching or moviegoing—and the researchers were forced to accept that the motivation to buy or choose a product has to exist before a subliminal ad has any effect.

There have been many attempts to harness the supposed power of subliminal advertising. The first was in 1957, when an ad man named James Vicary held a press conference at which he announced that his company had invented a new way of influencing consumers. He reported that he'd run a six-week experiment at the Lee Theater in Fort Lee, New Jersey, in which, throughout the hit movie *Picnic*, two alternating messages—"Drink Coca-Cola" and "Eat Popcorn"—had appeared for 1/3000 of a second, every five seconds. Vicary claimed that over the course of the test, forty-five thousand people were exposed to the messages, and he boasted that the marketing increased sales of Coca-Cola at the theater by 18.1 percent and sales of popcorn by an incredible 57.7 percent.

His announcement was a bombshell at a time when ideas like brainwashing and hypnotism were on everyone's mind. Unfortunately, the air was completely let out of the story when, in 1962, Vicary admitted that he had made up the entire study in an effort to drum up business for his marketing agency.

But that didn't stop others from claiming that they had developed their own methods of subliminally influencing people, notably the late Wilson Bryan Key. Key published the book *Subliminal Seduction* in 1973. The following year's paperback edition featured a great cover: a photo of a tumbler filled with a mixed drink, ice cubes, and a twist of lemon. It was a commonplace image, but the caption for the photo read, "Are you being sexually aroused by this picture?" Key claimed that thousands were. The book was a bestseller.

Did You Know . . . In 1978, the heavy metal band Judas Priest was accused of including a subliminal message in one of their songs. Groups of concerned citizens complained that the band's song "Better by you, Better than Me" included a hidden lyrical message saying, "do it." The band ignored the complaints, but then, in 1985, two young men in Reno, Nevada, attempted suicide. Their parents claimed the song made them do it, and they took the band to court. But when evidence was introduced that both boys had significant emotional problems and had talked about committing suicide, the case was dismissed. Wilson Bryan Key testified, but he did not verify whether there was any meaningful subliminal advertising in the song. He did, however, tell the court that subliminal messages could be found in Ritz crackers (he claimed the holes were arranged to spell the word "sex"), in the Sears catalog, and on NBC news.

Why do people faint when they see blood?

A lot of scientists will answer this question by simply saying "Nobody has a clue." That might be short and to the point, but it's far from satisfying. There are answers, but they're contentious, and one in particular has the uncanny ability to make some scientists' blood boil.

Lowered blood pressure, an irregular heartbeat, or low blood sugar can all result in a momentary loss of consciousness called fainting—or in medical terms, syncope (rhymes with "canopy"). Humans can faint for all sorts of odd reasons—you may hear of people fainting after coughing, after urinating, or after stretching. Sometimes, you can faint simply by getting out of a chair too quickly. In most cases, standing up suddenly causes blood to pool in the legs, lessening flow to the brain. If that dip in blood pressure is extreme, it knocks you out for a short time. You recover

because blood flow is re-established to your brain, either because you fall down or because you have the presence of mind to tuck your head between your legs just before you black out. In both cases, your head is positioned below, or at least no higher than, your heart, so refilling the brain with blood is easier.

But all these versions of fainting are straightforward physiological events triggered by physical stimuli, not mental conditions. How do we explain what happens when fainting is induced by pain, anxiety, emotional stress, or fear? Some people faint at the sight of a needle. We assume that's brought about by the expectation that pain will soon be inflicted, but that fear has little connection to the physiological consequences of, say, standing up too quickly. Instead of explaining these cases with physiology, we turn to evolution.

In 2005, Rolf Diehl, at the Krupp Hospital in Essen, Germany, suggested that fainting, with its accompanying drop in both blood pressure and heart rate, was a protective response exhibited by wounded animals. He reasoned that if an animal starts bleeding (and we are animals, remember!), its initial response is to constrict blood vessels and jack up blood pressure and heart rate, so as to maintain circulation in the face of blood loss. But if the bleeding doesn't stop and the animal's blood loss reaches a critical value—roughly a third of its total blood volume—the reverse kicks in: blood vessels loosen, the heart rate drops, and the animal's circulatory system slows until it eventually loses consciousness. (This isn't the same as playing dead. An animal playing dead still has its heart pumping and its nervous systems on high alert, exactly the opposite of being in a faint.) But while passing out may leave an animal vulnerable, Diehl argued that the drop in both blood pressure and heart rate buys precious time, allowing blood to clot and thereby reducing total blood loss. If the animal maintained a normal blood pressure in this critical situation, he argued, it would actually hasten death, not prevent it.

What's fascinating is that this straightforward physiological mechanism in humans happens not only in response to the loss of one's own blood but also to someone else losing blood. You'd think the emotion in play here would be fear—seeing someone else's blood raises fears that you'll see your own depleted next, and so your body shuts down and prompts clotting. But Diehl found that in cases where people fainted at the sight of blood, the feeling was one of disgust rather than fear. That feeling was followed by lowered blood pressure, slower heart rate, and passing out. If the body's response to heavy bleeding makes survival sense, the mind's response to disgust in these situations certainly does not—falling into a dead faint at the feet of a predator because you're disgusted is not a sensible survival strategy.

Did You Know . . . Human beings have a long history of reacting with disgust to bad smells, such as feces and rotting flesh. Not only that, our disgust reflex has evolved to be a reaction to unpopular politicians, unpunished criminals, and even a jacket once worn by Hitler.

The most interesting twist on this already peculiar reaction is that in the doctor's office, the disgust and fear that people feel is directed at the needle itself, not at the blood that it might draw. A needle might suggest the looming presence of blood, but really, most injections are close to 100 percent blood-free. So, something more must be at work.

One psychiatrist, Stefan Bracha, has proposed a controversial theory about needle fear. Bracha claims that this particular phobia dates back to Pleistocene times—as much as two hundred thousand years ago—an era he characterizes as extremely violent. Archaeologists have found definitive evidence that tribe-on-tribe and band-on-band disputes were frequent and gruesome. The weapons of choice were typically spears and axes. Deaths were gory and crude. The mere sight of blood was bad news. At some point in those conflicts, the best opportunity for survival—at least for noncombatants, such as children and young women—might have been to faint. A body lying in a heap on the ground could easily be overlooked, while one still upright remained a prime target.

A dig that began in 2012 on the shores of Lake Turkana in Kenya found evidence of a ten-thousand-year-old massacre, where it was obvious that sharp, pointed weapons like arrows and spears had been used in the slaughter. Of twelve almost-complete skeletons, ten had clearly died a violent death. Of course, this isn't proof of Bracha's thesis, but it does at least set the scene. If this fainting response has been passed down through generations, it might be genetic. At the same time, this hypothetical gene can't have been too widespread. Having entire tribes collapse on the ground at the sight of a spear would have been suicidal, if not genocidal. So if the gene exists at all, it is probably present in only a minority of people.

Bracha's argument assumes that this reaction is a human-only phenomenon, so the first time that a gorilla or chimp faints at the sight of a needle, the theory will have to be tossed out. So far, not surprisingly, no such experiment has been conducted. It's also true that there is more fainting when an expert (say, someone who's been doing it for thirty years) draws blood rather than someone less experienced. Apparently, the expert wastes no time coddling the patient and therefore appears to be more threatening than the less experienced practitioner.

Did You Know . . . Evidence suggests there's a hereditary component to fainting at the sight of a needle. More than half of patients who have a blood or needle phobia that prompts fainting have a parent or sibling who exhibits the same symptom. And the number of females who experience needle fainting is more than double that of males.

Theories like Bracha's make some scientists feel faintly uncomfortable. The broad field into which Bracha's theory falls is called evolutionary psychology, and its proponents have been accused of applying genetic explanations to modern-day phenomena when there's no justification for doing so. Take, for instance, the doubtful claim I once heard that the reason girls dream of monsters under the bed and boys dream of monsters coming in the window is that back in the days of our australopithecine ancestors, females roosted in trees at night while the males slept at the base; danger, therefore, came from different directions. There's zero evidence for any part of this claim, though it is entertaining. While Stefan Bracha's theories remain just that, he has at least applied evidence to a puzzle that otherwise lacks explanation.

Why do people choke under pressure?

Most of us associate choking—underperforming when success is expected—with sport. An athlete has a big game or competition and, despite years of training, when the moment comes for the star to shine, he or she fails—sometimes extraordinarily. Choking can happen in the classroom or the boardroom, too, but the sports examples seem the most vivid: baseball player Bill Buckner fumbling a routine grounder in 1986 or pro golfer Greg Norman at the Masters in 1996, leading by six strokes on the final round, then losing by seven in the end.

Psychologists tell us that as incentives for winning increase, people perform better—but only up to a point. After that point, performance starts to decline even as the incentives continue to rise. Two popular but opposed explanations are that choking is the result of either a distracted

mind or the opposite: a mind paying too much attention. A third possibility is a mind so excited, so on edge, that the nervous system cannot function at the level it must.

Let's use putting in golf as an example. Whether you putt at the golf club or at the mini-putt, you have probably experienced that moment when the pressure is on. Maybe a player has issued a challenge; maybe there's beer on the line. You have an easy shot—one you've done many times before. You prepare, you putt . . . and the ball stops three feet short of the hole. You haven't just missed—you've choked. You could have made that putt in your sleep!

Actually, that is the point of one of the theories: with actions that you have practiced and made automatic, interventions from your conscious mind hinder rather than help performance. Too much "top-down" brain activity is, at these very crucial times, the last thing you want. The unconscious mind has done the training and will serve you better. Too much focus can dismantle a familiar routine into too many pieces—like aiming, shifting, aiming again, rethinking when putting. One hint this might be a true picture of the situation is that experienced putters do better when they have less time for each putt (they've already developed an unconscious putting routine), whereas novices benefit from any extra time.

The alternative theory is that distraction, rather than overthinking, causes choking. Imagine extraneous information pouring into your brain. You have to deal with it along with the important task you're supposed to be fully focused on. Because you're doing two things at once, you fail at your main task.

The third theory suggests that overarousal and overstimulation, especially in extreme situations—such as competing for an Olympic medal—make people unable to perform at peak levels. Excess energy has nowhere to go, according to this theory, and the result is a lack of focus.

Sian Beilock at Barnard College has tried to make sense of the third theory by actually studying putters. She set out to see if putters were most likely to choke when they allowed their conscious mind to intrude at the very worst time. She worked with students, and although they weren't exactly pros, as part of the experiment they hit hundreds of practice putts, enough to make them sufficiently competitive. The students were then put into three groups. The first putted knowing there was money at stake, the second while forced to pay attention to distracting words, and the third in front of a video camera.

The group with money at stake and the distracted group each putted worse when faced with these challenges compared to putting without them. But the students putting while being videotaped did fine—their performance didn't decline with that challenge in place. That was interpreted to mean that these students were used to being recorded, and so they didn't have to think about being videotaped and avoided having it affect their performance. They were, in effect, inoculated against the pressure that leads to choking, whereas the others weren't.

One thing Beilock found in other studies was that expert players can detail for you the important parts of a good putting stroke, but they can't tell you nearly as much about the last putt they made, whereas novices are exactly the opposite. Psychologists call this expertise-induced amnesia. For experts, their skill is automatic, run by neural networks they're not even aware of. It's the difference between learning to drive and being an expert—after a while you don't think about the individual steps, you just do them.

A different study, which also observed that consciousness can induce performance failure, showed that if you count backward from one

hundred while you're executing a skill, like putting, you can occupy your conscious mind. It's too busy counting to mess you up in other ways, and so you survive the pressure and succeed.

It's also true that the process of becoming an expert in any activity is likely to change the physical brain. A nice example is juggling, an activity where thinking about what you're doing could be fatal if you're juggling fire sticks or chainsaws! With training, juggling becomes an unconscious routine, and at the same time, new neurons and connections to them are added to the motor control parts of the brain. Stop the training and the brain returns to its former state. But even with these additional brain cells, if unusual pressure is put on the juggler, other areas of the brain interfere, with unfortunate results.

Most of the thinking about choking involves differentiating between the conscious and the unconscious. Add up all the information flowing into your brain from your five senses, and the estimates are that maybe one-millionth of that information actually enters awareness. All the rest is unconscious. If we return to the example of driving, you'll see ample evidence of this: you don't think about the details of driving ("now check the rearview mirror, now look up ahead, now check my speed, now apply more pressure to the gas pedal"), you just drive. Your unconscious mind takes care of the subtle operations. Now put yourself on the putting green, with the distraction of cheering crowds, the clicks of cameras, and $100,000 at stake. Pretty hard not to overthink your putting stroke. But if you do . . . choke!

Why do I sometimes have difficulty recalling words on the tip of my tongue?

You are a rare person if you haven't experienced the tip-of-the-tongue phenomenon—that's when you're trying to remember a word or name and it feels right there, "at the tip of your tongue," but you can't quite access the information and spit it out. Words usually spill out of our minds and mouths at incredible speed without much mental effort, so it can feel awkward and frustrating when we suddenly can't recall the simplest thing. It's estimated that when you're in your twenties, you will experience this phenomenon about twice a week, but forty years later that rate will have doubled.

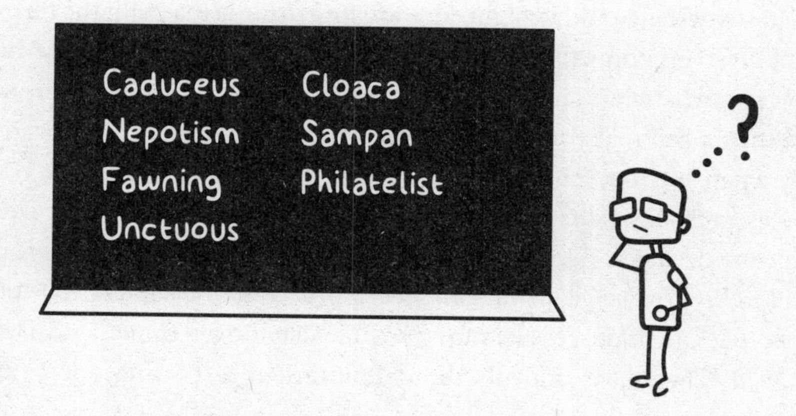

If you've experienced this phenomenon, you can attest that it provides a small, short-lived window into your own brain. It's like waving a flashlight around inside your head, trying to illuminate a word that you know is waiting there but that is playing a game of hide-and-seek. And not only do you know that pesky word is in there, you even have a sense of its size and shape. But you still can't find it.

In 1893, the great American psychologist William James noted that when we're searching for a word and the wrong ones are offered to us, we know that they're not what we're looking for, that they don't fit the mold. But that doesn't solve the problem.

So why are psychologists like James interested in this experience? Because it's like seeing a hummingbird's wings in slow motion: an opportunity—maybe—to study the processes of memory and word-making more closely. The next time it happens to you, take a moment to savor the feeling and explore it instead of pulling out your hair and fearing for your sanity.

TRY THIS: Want to experience tip-of-the-tongue effect right now? Try naming all seven dwarfs and watch what happens.

Here's what you might find: the difference between a tip-of-the-tongue experience and straight-forward forgetting is that with the former, you know you know the word you're looking for. It fits the sentence and expresses what you want to say. In that sense, it's much closer to the surface than a fact that you have completely forgotten and become aware of only when you're reminded.

Research shows that you likely know a lot about that lost word even if you can't dredge it up. The classic tip-of-the-tongue study that revealed this (and set the stage for the modern approach to the subject) was published back in 1966, by Harvard psychologists Roger Brown and David McNeill. They gave students the definitions of forty-nine uncommon words, defined as occurring less than once in a million words in common usage. (They didn't include definitions of "super-rare" words, ones you come across every four million words.) They did not supply the words, just the definitions. (Whether these words are as common today, more than fifty years later, is arguable, given that they included *ambergris, caduceus, nepotism, fawning, unctuous, cloaca, sampan*, and *philatelist*. Don't worry: if you don't know the meaning of most of these, you are not alone!)

The psychologists then asked the students to come up with the right words to match the definitions. And the students came through, experiencing countless tip-of-the-tongue moments, memorably described by Brown and McNeill as appearing to be in "mild torment, something like the brink of a sneeze." When students thought they knew the word but couldn't immediately verbalize it, Brown and McNeill asked them to list any features they thought they knew, such as the number of syllables, the first letter, and words of similar sound and similar meaning. The intriguing thing was that the students were able to do quite a bit of that.

Guesses at the number of syllables were impressively accurate. For instance, when people were searching for three-syllable words, two-thirds of them guessed the mystery word was indeed three syllables. The same accuracy rate held true for shorter words, but it was lost for words longer than three syllables.

Sometimes the students were seeking the wrong word, but when they were right, they were able to guess the first letter accurately 57 percent of the time. For example, when the students were told the word was defined as a flat-bottomed boat common in Asia and propelled by oars (a sampan), those experiencing tip-of-the-tongue effect recalled similar-sounding words like *Saipan, Siam, Cheyenne, sarong, sanching*, and *sympoon*. (The last two are not actual words, by the way.) The students were often able to come up with words with similar meanings, such as *barge, houseboat,* and *junk*—all of which make sense but are more common words and are not physically or structurally like *sampan*.

Perhaps the most surprising result of this experiment was that when words of six letters or more were the target, students did well at guessing letters at the beginning and end of the word, but poorly when it came to letters in the middle. This led Brown and McNeill to the image of the mystery word as the tall man in the bathtub: his head sticks out at one end, his feet the other, but his middle can't be seen.

Explaining these results in terms of how words are stored in our brains is tricky. Brown and McNeill argued that, yes, words are stored in the brain in their entirety but can be retrieved without having to know every single letter. That was made clear by the students' correctly guessing the first and last letters in the experiment, but it's also true when y—— r——d th— l—st f—w w—rds —f th—s s—nt—nc—.

The reason you feel the tip-of-the-tongue phenomenon so strongly is that you are aware of the meaning and appropriateness of the word but just can't connect those things to the actual letters. It might happen because you don't use the word often (or haven't used it for a while) and so the connection has gradually weakened. And age is a factor, too.

Since this classic 1966 study, there have been dozens of attempts to clarify what sorts of cues might help resolve a tip-of-the-tongue quandary. One of the most interesting was conducted by Lisa Abrams and Emily Rodriguez in 2005. They were interested in what sorts of words would help relieve the tip-of-the-tongue state. So with the question "What do you call a large colored handkerchief usually worn around the neck or head?" they were looking for the word *bandana*. If one of their subjects ran into a tip-of-the-tongue issue, he or she was asked to read

from one of three word lists. One list included a word with the same first syllable, such as *banjo*; the other list included a word that had the same first syllable but was a different part of speech, such as *banish*; the third list had unrelated words.

What Abrams and Rodriguez found was that *banish* helped but *banjo* didn't. Why? They concluded that words that are the same part of speech as the target word (in this case, the two nouns *bandana* and *banjo*) compete in the brain, whereas *banish*, a different part of speech stored in a different part of the brain, helped recall.

Did You Know . . . Synesthetes—people whose sensory experiences cross wires—can hear colors or taste words. When synesthetes experience tip-of-the-tongue effect, they report tasting the word they're looking for before they know what it is they are tasting!

One very cool new addition to this study is an online diary where you can record your tip-of-the-tongue experiences, which are then made available to researchers. Spedi—or Speech Error Diary—is run out of the University of Kansas. Citizen science at its best!

Why can't I remember anything that happened before I was two years old?

You may already be thinking, "But wait! I distinctly remember my first birthday!" or "My first trip to Niagara Falls was when I was one and a half!" But those memories could easily have been created by seeing photographs or hearing events described by others, over and over, not necessarily by remembering the actual experience. Still, there are some exceptional people who really do remember things from a very young age. But no more than 1 to 2 percent of the population seem to be able to do that.

For the vast majority of people, first memories can be recalled from the age of about three to three and a half years. A child's brain is rapidly

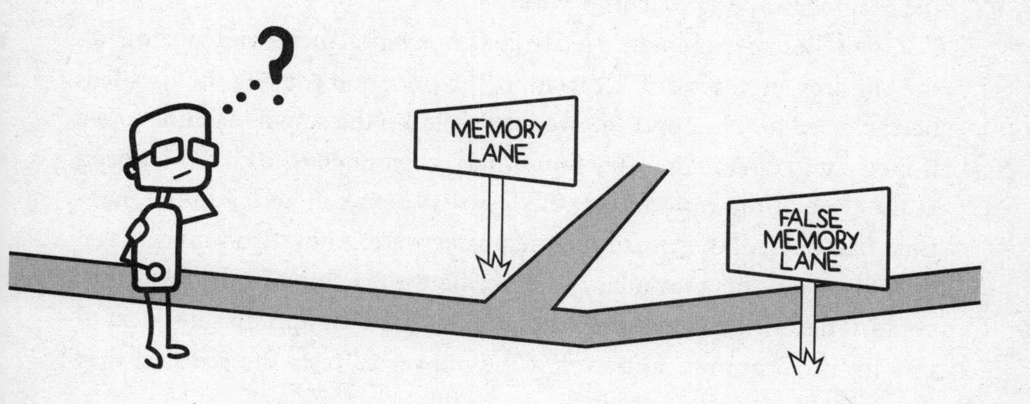

developing before that, and somehow, during that crucial process, stored memories seem to mysteriously vanish.

Researchers have tried to prove that children under the age of two can actually store memories, but because very young children lack a good command of language, proof is difficult. But New Zealand psychologists Harlene Hayne and Gabrielle Simcock found a way around this problem, by building something called the Magic Shrinking Machine.

The Magic Shrinking Machine was used with children ranging in age from just over two years old to just over three. Each child learned to start the machine by pulling a lever that turned on lights. One of the experimenters then put a large toy into the machine, making it "disappear," then the same experimenter turned a handle to produce a set of sounds. Finally, the child was shown how to retrieve the toy from the machine. Lo and behold, when the child retrieved the toy, it had shrunk (or at least appeared to have shrunk)! The child repeated the actions seven times, each time with a different toy that magically shrunk. By the end of the experiment, the children were able to repeat the entire sequence by themselves. This proved they clearly remembered what to do.

Six months later, then a year later, the children were tested to see if they could remember the Magic Shrinking Machine and how it operated. They were tested using two different methods. One was verbal: they were asked questions about the machine, like "Last time I visited you, we played a really exciting game! Tell me everything that you can remember about the game. What were the names of the toys? And how did we make the Magic Shrinking Machine work?"

Then the experimenters tried to coax nonverbal memories by showing the children pictures of the machine, the toys, and the bag the toys had been carried in. They even showed the children the actual machine to see if they could operate it. They found that the younger the child had been at the time of the experiment, the less they remembered, and the more time had passed, the worse the memories were. They also found that if a child lacked the vocabulary to describe the machine in the past, they couldn't describe it later—even if they'd acquired the appropriate vocabulary in the meantime, and even if the nonverbal tests showed that they remembered the machine in some detail. This suggests that our ability to talk about our early memories is limited by our ability to process language.

Six years later the children were tested yet again. Many remembered the machine, even those who were only two years old at the time they'd first been introduced to it. That finding was surprising. It showed that the language barrier wasn't as impenetrable as once thought. But it was still significant for a lot of test subjects.

A number of theories have been proposed to explain what is necessary for us to form crystal-clear memories. One is that we can't do that before we realize we have an identity. That happens sometime around the age of two—roughly the same time that we start speaking and understanding words. It's not yet understood how those two events relate, but there is evidence that the kinds of conversations children have with their parents play an important role in establishing memories, especially those that can be described verbally.

Did You Know . . . A maître d' from a restaurant in Washington, D.C., claimed that the second time you visited his restaurant he could recite exactly what you ate the first time. He was also certain he could remember events from the first year of his life. But there was no way of verifying his childhood memories.

Of course some events in a child's life much earlier than this—even in the first few months—can still influence emotional life into adulthood, especially if those events are extremely stressful or traumatic. But these "memories" are not actively remembered. The youthful developing brain is likely to be more forgetful, too. It's busy rapidly assembling the structures and networks necessary for laying down permanent memories. And the creation of a memory involves multiple tricky steps: recording the memory, stabilizing it, and preparing it for storage in long-term memory banks. Making it through each of these steps is a risky process: if a memory isn't firmly implanted, it will be lost along the way, forever forgotten.

Did You Know . . . Ultimately, forgetting some of what we experience is necessary: those rare humans who have the (dis)ability of remembering virtually everything are often, if not always, miserable. A woman known only as A.J. (for her privacy) has an extraordinary memory for dates and is not happy about it: "Most have called it a gift but I call it a burden. I run my entire life through my head every day and it drives me crazy!!"

So when you think back to the earliest events you can remember, the sketchiness of those few memories you've hung on to is likely due to the fact that your two-year-old brain was busy frantically building an effective memory system. Once you turned five or six, that system was pretty much in place, and your detailed reminiscences—true memories—began to be recorded from that time on.

Why does time seem to speed up as we age?

There's no doubt that the vast majority of people feel that time moves faster as they age, but very few of them bother to estimate by how much.

A century ago the great American psychologist William James suggested that as we grow older, and more jaded and worldly, we enjoy fewer remarkable experiences in a year, and so the years become less and less distinct from each other. Another theory suggests that because each successive year is a smaller percentage of one's overall life, it is less significant when weighed against the rest and therefore passes by virtually unnoticed. When you were ten, every year was huge: 10 percent of your life. At age forty, though, one year is only 2.5 percent of your total life.

There's also a phenomenon called forward telescoping. Imagine you're asked when you last saw your aunt and you say, "Uh . . . three years ago?" when it's actually eight years since you saw her. You've zoomed in time, bringing the past closer than it really is. When someone asks me how long ago an event took place, I double my first estimate, and even then I sometimes underestimate the passage of time. That's forward telescoping.

Sometimes I feel time is catching up to me.

In the mid-1970s (remember how slowly time passed then?), Robert Lemlich of the University of Cincinnati proposed one significant adjustment to the idea of the apparent passage of time versus reality. He argued that since time is all subjective anyway, years are also subjective. Calculating what percentage of your total life is represented by each passing year is fine, but it's strictly mathematical and so doesn't take into account that each passing year feels shorter as well—it is a smaller part of your total life numerically, but it feels even less than that. It's all in your head, really, so your estimate of the length of a year that has just passed should be compared not to how long you've lived but to your sense of how long you've lived.

Lemlich created equations to quantify what he meant. Their implications are surprising, even shocking. Let's assume you are a forty-year-old. Lemlich calculated that time would seem to be passing by twice as fast now as it did when you were ten. (Remember how long summer vacation seemed to last?)

But there's more: the numbers tell you that if you're that forty-year-old and you're going to live to eighty, you're halfway through your life by the calendar, but because time seems to be passing ever more rapidly, Lemlich's math suggests you will feel you have less time left than you actually do. By his calculations, at age forty, you have already lived—subjectively—71 percent of your life. It gets worse: by the time you're sixty, even though you have twenty years remaining, those twenty years will feel like a mere 13 percent of your life.

These numbers are shocking enough, but they take on an even more bizarre twist when you extrapolate them back and ask the question: At what point in our lives have we experienced half of our subjective life? If you're that forty-year-old, you will have experienced half your total subjective life by the time you were twenty. Even if you live to a hundred, 50 percent of your total life experience will feel locked in by your twentieth birthday.

Lemlich backed up his numbers with experiments. He asked a group of students and adults to estimate how much slower time seemed to have passed when they were either half or one-quarter their present age. His theory predicted the answers almost exactly: time seemed to have passed only half as fast when they were one-quarter their present age, and about two-thirds as fast when they were half their present age.

Is something else going on in our brains that would change our perception of the passage of time as we age? It might be that our internal clock (and jet lag and shift work demonstrate just how crucial that clock is) runs slower as we age. If your clock now estimates a minute to be three minutes, because it's running slower, then many more events will be packed into that time frame and it will seem that time is passing faster.

An extreme example is the case of a man who, at the age of sixty-six, was admitted to hospital in Düsseldorf. Examination revealed a tumor in the left frontal lobe of his brain. He'd gone to the hospital because he was finding life unbearable: everything was happening at breakneck speed. He had to stop his car by the side of the road because the traffic was too fast. The television, already manic, was triple manic, and as a result of this experience, he had begun to withdraw from society. When asked to estimate the passage of sixty seconds, it took him four and a half minutes. Imagine what traffic would look like if four minutes' worth was packed into a minute! What this case suggests is that disruptions to certain parts of the brain alter our perception of the passing of time, and while this particular case was unusual, it's possible that a gradual and minor version of this affects everyone's sense of time passing.

TRY THIS: Want to get a sense of your subjective age? It's simple: estimate how old you will be at death. Divide that number into your current age. That gives you the percentage you've lived of your chronological age. But if you then take the square root of that, you have the percentage you've lived of your subjective age. Be ready to be shocked! Here are my numbers. I come from long-lived parents, so I should live to ninety. That's twenty more years—yay! But that twenty represents only 11 percent of my entire life experience. That means that I have already experienced close to 90 percent of my subjective life. Boo.

You might be wondering why we're spending time (it's precious!) figuring out equations to account for how we experience time. This kind of data supports what might otherwise seem to be mere impressions like this one by Robert Southey, the poet laureate of England in 1837: "Live as long as you may, the first twenty years are the longest half of your life. They appear so while they are passing; they seem to have been so when we look back on them; and they take up more room in our memory than all the years that succeed them."

TRY THIS: The next time January rolls around, pay attention to the number of times you date anything and write the previous year. I used to do that consistently, almost to the end of January, but in the last two years I haven't once written the incorrect year. I have no idea why. Keep an eye on it yourself. Report back to me what you learn!

Can we really tell when someone is staring at us?

Have you had this experience? You suddenly feel that someone is staring at you from behind. You turn to check and they are in fact looking right at you. How could you have known they were doing that?

For centuries, people have believed they can feel stares before they see them, but it wasn't until about a hundred years ago that anyone attempted to explain this through the lens of science rather than magic or parapsychology. In 1898, British psychologist Edward Titchener presented an explanation that is still favored today. He began by observing that people are concerned about being viewed from behind, and will, if seated at the front of a crowded auditorium, constantly adjust their hair or brush their collar or even glance behind them. In turn, those movements attract the

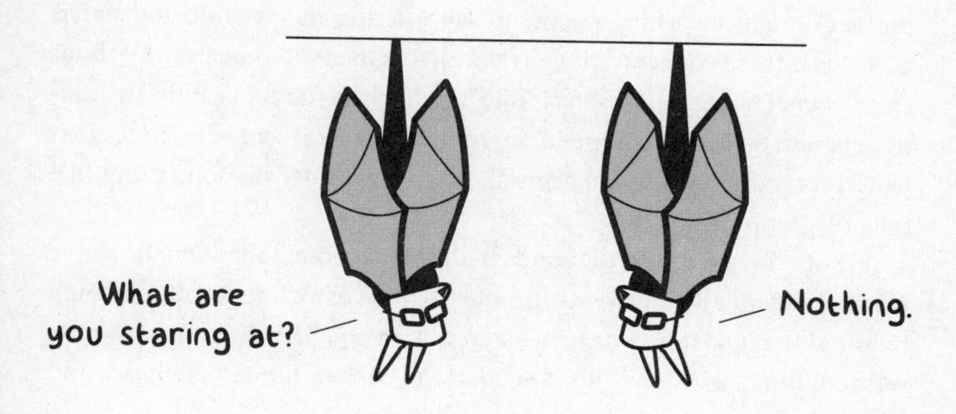

attention of people behind them. So when the person in front happens to look back, she will inevitably see someone looking at her, and will be convinced she sensed that stare. Titchener's theory cleverly turned the situation on its head (literally). And the timing works: while it takes about a second to turn your head, it takes the starer only about a fifth of a second to shift his or her gaze to you.

In 1913, John Edgar Coover followed up on Titchener's work. He had a "starer" stare at a person based on a roll of a die. When the starer rolled odd, he was permitted to gawk at a "staree." When he rolled even, no gawking. When he stared, he was to stare for fifteen seconds straight. Meanwhile, the staree was tasked with guessing, without turning around, when he was being stared at and when not. Coover's results showed that the person being stared at could not accurately guess when it was happening. The guesses—at chance levels, 50.2 percent—were no better than results obtained by simply flipping a coin.

In this study and others, some issues continue to challenge clear answers to the question. Did the starer randomize the timing of the stares? If not, then positive results could be the result of the staree starting to anticipate when he was being stared at by sensing a rhythm. The staring had to happen at random intervals for the test to yield accurate results. Nor could there be any sounds involved—no shifting in the chair, no rustling of a shirt, no clues whatsoever of the starer's gaze.

In 1993 at the Institute of Transpersonal Psychology in California, William Braud conducted some eye-opening research. Braud felt that the problem might be asking people to say whether they were being stared at. Their intense concentration on stares, together with uncertainty about exactly what being stared at feels like, might blind them to subtle, instinctive signals. So Braud equipped his volunteer starees with electrodes that would record physiological arousal. This apparatus was something like the polygraph.

He put the starer and the staree in different rooms, allowing the staree to simply sit quietly for twenty minutes in front of a camera without being required to guess if he was being stared at or not. The starer, meanwhile, was looking at a TV monitor aimed at the back of the target's head, and he or she either stared or not, depending on cues given. The routine was ten stares, ten non-stares, each lasting thirty seconds, randomly ordered.

The results were sensational. There was a significant timing correlation between strong physiological responses in the staree and the actual stares from the person in the other room. Not only that: there was strange evidence of the two people being connected physiologically. As the starer became more comfortable with the odd situation of staring at the backs of people's heads, the starees' arousal levels declined. All of this, remember, with two people in separate rooms with a camera and a monitor. It boggles the mind.

As convincing as it seemed, this experiment is not the final word on the existence of a stare-guessing instinct. A similar set of experiments was conducted in the mid-1990s by the two-person team of Marilyn Schlitz, a well-known believer in psychic phenomena, and Richard Wiseman, a well-known non-believer. To their credit, they decided to partner up for the experiments. Using a setup similar to Braud's, they performed tests in Wiseman's lab in England and in Schlitz's lab in California. All were done the same way, but astonishingly, when Schlitz was the starer, the staree (whoever it was) predicted stares at a rate far exceeding the rate of chance. However, when Wiseman was the starer, the staree predicted at rates no better than chance. Both Schlitz and Wiseman were confounded by these results. So they took the next step: they designed an experiment in which they would switch roles. One would greet the staree at the beginning of the experiment, while the other would do the staring; then they'd switch. The justification for this was the admittedly slim possibility

that Marilyn Schlitz's high results were somehow due to her establishing better rapport when she met the starees. In this new round of tests, she would sometimes meet the staree and sometimes not.

The experiment failed to find any evidence of psychic staring, no matter who stared or who greeted. This left the team with the difficult situation of having conducted almost identical experiments three times, some with at least partial positive results (Marilyn's), and now once with negative results for both Schlitz and Wiseman.

Allow me now to back up a bit, to the 1980s and renegade British scientist Rupert Sheldrake, who stared deeply into the heart of the problem. He is best described as anti-establishment, arguing for phenomena that, according to orthodox science, couldn't happen. Knowing you're being stared at was a phenomenon he set out to prove was real, and he even speculated on how it would work.

Sheldrake argued that the explanation lies in the mechanics of vision. Science sees vision as a sequence: light enters our eyes, triggering neural signals in the brain, and the patterns of light are analyzed to form images. Light enters the eye—nothing leaves it. That's what every single piece of evidence about seeing supports. But Sheldrake argued that this is only half the picture that our eyes also send out unperceived signals or "fields," and these affect what the eyes are looking at. There is only one problem with Sheldrake's theory: there isn't a shred of evidence to support it.

The word "field" can be taken to mean just about anything, and because science is full of fields—gravitational, electromagnetic—the word has the ring of truth about it. Until there's real, solid evidence of "psychic" staring, in a controlled environment yielding consistent results, the idea of stare prediction remains an intriguing idea only. The eyes may have it, but so far, we have no proof.

What are near-death experiences?

It's easy to describe what the typical near-death experience (NDE) is like, but so far it's been impossible to explain exactly what it is.

It's not uncommon that someone who is either severely ill or gravely injured comes to a point where they're close to dying: their heart has stopped, and electrical activity in the person's brain has flatlined. If that situation were to persist, it would be death. But there are many who have pulled back from that brink and continued living a normal life.

Of those, a small number report having had a near-death experience. That could involve an out-of-the-body experience: hovering over the operating table looking down on their unconscious body, or traveling

down a tunnel with a light at the end of it, meeting deceased family and friends, feeling at one with the universe or encountering some sort of spiritual being. There have also been unpleasant NDEs—the feeling of complete nothingness or being in a place peopled by demons and threatening animals—but those are much rarer than the positive ones.

What could be happening to create such vivid images in a person who's near death? It comes down to this: if a person has an NDE when their brain activity is zero, it suggests that mental activity can happen independently of the brain. That idea is embraced by many—it's a standard part of the belief in a "soul"—but scientists are convinced that our mental life is generated by the brain and exists nowhere else.

In 2001 a Dutch scientific research team reported in the medical journal the *Lancet* their studies of NDEs experienced by cardiac-arrest patients. They identified a short period of time during which the heart had stopped and the brain presumably flatlined as the only possible time when these patients could have had their experiences. The researchers asked: How could a clear consciousness outside one's body be experienced at the moment that the brain no longer functioned?

Fewer than 10 percent of the patients they studied had a substantial NDE, but the details reported by those who did were eerie: one man remembered specific things doctors were doing around him, including removing his dentures. He even sensed the pessimism of the medical team.

Reaction to the Dutch team's findings from other scientists was predictable. An editorial in the same issue of the *Lancet* argued that it was

difficult to prove that the patients' NDEs happened exactly as their brains were flatlined as opposed to just before or after. It also raised questions about the veracity of the reports, because several people who originally said they hadn't had an NDE changed their minds when asked again two years later. The Dutch team argued that because the majority of their patients hadn't experienced an NDE, the standard scientific explanations of the brain responding to high levels of carbon dioxide or low levels of oxygen didn't make sense. Otherwise, virtually all patients would have had those experiences.

The doubts about the 2001 study focused the challenge sharply: When there is sure evidence that the brain has flatlined, is there ever any mental activity? And how can the stories of people claiming to have had NDEs be confirmed?

Did You Know . . . In 1977 a woman named Maria in a Seattle hospital suffered cardiac arrest. She recovered, and the next day she reported having seen herself above the operating table, then looking outside and seeing a distinctive tennis shoe on the window ledge. The shoe was located and the details fit Maria's description. This seemed to suggest an NDE that involved an out-of-the-body experience. But later two skeptics returned to the hospital and placed a shoe on the same ledge. They noted two things: first, the shoe was easily seen from the ground and could well have been mentioned in Maria's presence. But even more damning was that every detail of the shoe could be seen from inside the hospital room, even from the bed.

A recent study took advantage of the fact that many NDEs include an out-of-the-body experience, usually of the patients looking down on themselves and the medical teams. For these to happen at all is almost unbelievable, let alone for them to happen during cardiac arrest! In this case the researchers installed shelves in the operating rooms where emergency procedures were likely to happen, then put objects on those shelves that could be seen only by someone floating in midair.

In more than 2,000 cases of hospital admission for cardiac arrest, there were only 330 survivors and only a handful of interesting cases of NDEs. In the end, only one patient was able to describe verifiable events while he was in cardiac arrest. Sadly, there were zero instances of patients describing anything hidden on the shelves.

There are occasionally small pieces of research that show that there's more to discover. One Canadian study in early 2017 found that of four patients who had life support withdrawn, one continued to show peculiar brain wave activity for ten minutes after the heart stopped. It was only a case of one, but it was completely unexpected. And another recent study in rats showed intense activity in the brain immediately after cardiac arrest—a burst that resembled, and even exceeded the brain activity seen in awake rats. That also came as a surprise. Was this activity due to a shock to the brain, or was it rodent enlightenment?

The frustrating thing about the divide between believers in NDEs and nonbelievers is that the controversy sits right at the heart of one of science's most challenging puzzles: consciousness. What goes on in our brains to generate the thoughts, dreams, ideas, and images that we experience when we're conscious? Is the brain necessary for those things? Many who believe in NDEs think not; scientists disagree.

What is déjà vu?

Chances are you've experienced déjà vu before, in which case, one thing is for sure: you remember what it was like. Ironic, because the essence of déjà vu is a failure to remember. Déjà vu is that sudden vivid feeling that wherever you are and whatever you're doing, you've been there and experienced that situation before, even though you have no actual memory of it.

Technically, there are several slightly different versions of déjà vu, including déjà vécu (the feeling of having lived through an event), déjà senti (the feeling of having already felt a particular emotion), déjà visité (the feeling of knowing a place you've never visited before), and déjà pensée (the feeling of having had the same thought before). (Do writers have déjà écrivée?) Despite the shades of gray, each variation is really about the same sort of experience. Consider this example of déjà vu. You walk into a room full of people, and as your attention is pulled one way, then another, you briefly notice a lamp on a table.

Later, as the crowd thins, you get a better look at the lamp and think, "I've seen that before," which expands to overwhelm you with the feeling that you've actually been in that room before. It might be only a momentary occurrence—as that instant feeling of familiarity fades quickly—often because you're busy trying to remember why it's happening in the first place. You think, "Where was I when something like this happened previously? What is it that's so familiar about this situation?" Rationally, you know that you haven't really been in that situation before, but something's going on in your brain to make you think you have.

Déjà vu is a frustrating psychological occurrence. It is estimated that roughly two out of every three people have experienced it. Despite how common it is, déjà vu is poorly understood. Even the experts—psychologists, neuroscientists, psychiatrists—acknowledge that they don't have a precise handle on it, but they would also say that they're getting closer, dispensing with old ideas and refining new ones.

Given the lack of an exact scientific explanation, it isn't surprising that we have poets and authors to thank for many of the most vivid descriptions of the déjà vu experience, as well as some of its worst explanations. In the "best description" category, here's David Copperfield, in Charles Dickens's classic tale: "We have all some experience of a feeling, that comes over us occasionally, of what we are saying and doing having been said and done before, in a remote time—of our having been surrounded, dim ages ago, by the same faces, objects, and circumstances—of our knowing perfectly what will be said next, as if we suddenly remembered it!"

If you've experienced déjà vu, you'll realize that Dickens's observations are right, but it's the part about knowing what will come next that's really intriguing. How many people actually feel that level of clairvoyance? And how long does that feeling last? If someone says something and you think you know what he or she is going to utter next, then you should be able to predict the next statement. But there is no proof that the feeling of a prediction leads to actual predictions.

Dickens's reference to "dim ages ago" may be, as countless others have suggested when it comes to déjà vu, evidence of reincarnation. Of course, reincarnation would be a simple, straightforward explanation . . . if it weren't for two major stumbling blocks. First, there's no credible proof that reincarnation is possible. And second, if one did previously experience an identical situation "dim ages ago," how likely is it that it would be remembered?

Think of a conversation you could have had in your living room today. Then imagine a previous life in which you might have been sitting in front of a campfire, dressed in your filthy, lice-ridden woolens, or if you were of higher social standing, sitting in front of the fireplace dressed in your fashionable, lice-ridden woolens. It's unlikely that the conversation you had in the dim past would have been about binge-watching *Breaking Bad*. Rather than reincarnation being the explanation for déjà vu, perhaps déjà vu led to the very idea of reincarnation in the first place.

Other paranormal explanations run through déjà vu. In the words of "thought energy expert" Jeffry R. Palmer: "Far from discounting the study of the paranormal, the recent theories describing déjà vu experiences as electro-chemical misfiring in the brain . . . highlight the importance of continued research into paranormal phenomena."

Actually, "electro-chemical misfiring" isn't a justification for researching the paranormal; it simply means that there's something odd going on in the brain. The question remains: What is that something?

Nearly a hundred years ago, psychologist Edward Titchener suggested that déjà vu was a result of a broken sequence of thoughts. Imagine, he suggested, you look both ways before crossing the street and just for a second your attention is caught by the display in a store window across the road. You then cross the street, but when you catch a second glimpse of the store window, you think, "I've been on this street before." Titchener's argument was that the original left-right look is separated in your mind from the image of the store window by the actual crossing, which makes you think that the first experience (the looking) is a much earlier and fuller experience in the past than it actually was.

Another possibility is that the processes of memory making and memory retrieval, which most of the time are independent, occasionally happen simultaneously, resulting in a feeling of remembering a situation you're actually experiencing for the first time. This was once likened to having the record and playback heads of a tape recorder active at the same time. But the tape recorder analogy isn't a great one anymore (fewer and fewer people even know what one is!), and the theory isn't that popular, either.

There's another explanation that is the opposite: this one relies on two normally synchronous processes separating momentarily. One example—and this is a hypothesis only—is that a sense of familiarity and the retrieval of the memory associated with that familiar situation usually happen at the same time. But if the timing is disturbed and the feeling of familiarity jumps ahead of the memory, you're left thinking, "I know this place," with no relevant memory to support the feeling. Result: déjà vu.

Finally, another explanation hinges on features of a place that trigger the déjà vu experience. Imagine you walk into a room in which you find an old television exactly like the one your grandfather owned. The TV is familiar, but it's out of context, so you can't put your finger exactly on where you've seen it before. Instead, you think you've been in the room before.

TRY THIS: I'm sure that you have driven some distance on a highway either without remembering it because you were listening to the radio or, conversely, without hearing what's on the radio because you were concentrating on driving. That latter situation is a perfect opportunity for déjà vu. You've heard what was said, but you didn't notice it or weren't conscious of it. If you were to hear one of them again, it's possible that it could be dredged up from wherever it landed in your brain. It's quite likely it's there somewhere, and there's a lot of evidence that these unconscious thoughts influence us all the time. Try to remember what songs you heard on the radio in your last car trip. Can you remember them all?

Although specific objects can trigger déjà vu, the experience is not necessarily tied to or triggered by objects. Anne Cleary at Colorado State University has done many experiments showing that déjà vu can be triggered in the lab. For one, she created two sets of pictures that had similar layouts but very different subjects and details. She started by showing student volunteers one set of images. When she showed them the second round of images with the structurally similar scenes the students were convinced they'd seen the pictured place before. It turned out that it was the configuration of the items in the picture, not the specifics of those items, that stood out in the students' minds and led to the déjà vu. Dr. Cleary has extended these experiments by using virtual reality technology, and when students are immersed in the scenes presented to them, the déjà vu feeling can still be created. The experience is called feature-matching, and Cleary says it can "produce familiarity and déjà vu when recall fails."

Of course these experiments are at best only a partial explanation of déjà vu. The composition of a scene is one thing, but a real-life, complete experience includes people, conversations, and many other sensory details, none of which are included in Cleary's experiments. Psychologist Bennett Schwartz described experiencing Cleary's feature-matching. He had a powerful déjà vu while touring a castle in Scotland. When he was in the gift shop after the tour, he saw photos from a movie shot in the castle. He then recalled that he had watched that film five years before. Presumably, it was the memory of the movie scenes that triggered his déjà vu.

Science Fact! *For some reason, castles are excellent places for déjà vu. Even before Schwartz, at least five déjà vu experiences documented in the literature happened in castles—English, Scottish, and German—from the 1860s to the 1950s. Do we all have vivid mental images of castles even if we've never been to one—images that could set off a déjà vu?*

So far, there's no hard data to support any one explanation of déjà vu. Something is going on in the brain, and although the specific nature of that glitch hasn't been proven, some neurological conditions are associated with a much greater frequency of déjà vu. One in particular is temporal lobe epilepsy. Many sufferers experience déjà vu shortly before they have a seizure. Temporal lobe seizures are usually the result of a small area of scarring in the brain beside and behind the ear. It has been suggested that small disruptions in that region might produce the déjà vu effect, even in people who don't have epilepsy.

Some researchers have tried to determine who among the general population is prone to déjà vu. It's extremely rare in children below the age of eight or nine, but its incidence rises rapidly through childhood before starting to decline later in adulthood. And it seems to happen more often when people are stressed, tired, or anxious. In 2014, researchers reported a remarkable case of a twenty-three-year-old man who experienced déjà vu as a result of chronic anxiety. His case was so severe that he stopped watching TV and reading newspapers because he felt he had seen and read everything already.

A handful of other people have reported such overpowering déjà vu. One woman stopped playing tennis because she felt she could anticipate the result of every rally. The apparent champion of déjà vu frequency is an otherwise mentally healthy individual called Mr. Leeds. In the 1940s he tracked his déjà vus for a year and reported 144 of them—all in incredible detail! Still, many of these are extreme cases accompanied by other psychological disturbances.

Did You Know . . . Educated people who travel are more prone to déjà vu. It's possible this is because they've stored more images and memories that resemble new encounters. Also, dreamers, especially those who remember their dreams, are more likely to have déjà vu.

We can't leave this subject without acknowledging déjà vu's evil twin: jamais vu, the feeling that you've never been in a place that you're very familiar with. It's been suggested that jamais vu is like "word blindness," where if you read or speak a word over and over, it starts to become meaningless, to lose all familiarity. Take the famous *Friends* episode called "The One with the Stoned Guy." Jon Lovitz, who is playing a restaurateur in search of a new chef, arrives at Monica's apartment to sample her cooking. It's clear to the others in the room that Lovitz's character is high, and when Monica mentions the tartlets, Lovitz excitedly repeats the word three times, as though he has something else to say on the matter, before announcing, "The word has lost all meaning." If you take that sort of mindless repetition that's devoid of meaning and multiply it by ten, you get closer to the experience of word blindness, or roughly, jamais vu.

History Mystery

Did Archimedes really set Roman ships on fire with the sun?

This story isn't as well known as the one about Galileo dropping balls from the Tower of Pisa or Newton sitting under the apple tree, but it is more dramatic, more fabulous, and even more contentious.

In 213 BC, Rome was at war with Carthage, the seafaring nation operating out of what is now Tunisia. The island of Sicily was largely occupied by Rome, but the city of Syracuse on the island was controlled by Carthage. The Roman army and navy attacked the city, but it was well fortified with a walled harbor, and the Greek genius Archimedes was overseeing its military. He was undoubtedly one of the greatest scientist-mathematician-engineers of the time, if not of all time.

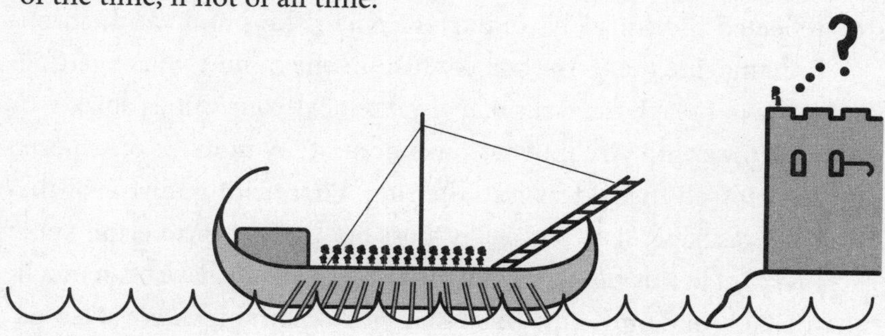

Archimedes's exploits in 213 BC were more spectacular. The Romans had giant oared ships armed with devices called sambucae—huge ladders with tons of rock counterweights at the bottom. These could be swung up to rest against the harbor's walls, allowing the soldiers to scramble over the ramparts and enter the city.

Science Fiction! *Archimedes is best known for the (likely apocryphal) story of jumping out of his bathtub naked and running through the streets yelling "Eureka!" after he figured out that the volume of water displaced must be equal to the volume of the part of his body that was submerged in the tub. Archimedes's insight led him to the solution to the problem of determining if the king's gold crown had been adulterated with silver. His experience in the bath meant he could measure volume, compare it with weight and determine an object's density, an important indicator of purity (as gold is nearly twice as dense as silver).*

If the Romans thought sambucae gave them the technological edge, they were sorely mistaken. Archimedes had set up a variety of devices to ward off the ships, the most remarkable of which was (if the story is true) a giant mirror (or set of mirrors) that reflected the sun's rays onto the Roman ships and set them on fire. Imagine the terror and confusion: one minute you're getting ready to breach the walls of a city, the next your ship is ablaze.

How could Archimedes have done it? A number of experts have tossed their ideas into the ring. First, with a mirror of the correct shape, it is physically possible to concentrate the sun's rays. But in this case, a single mirror would likely have been much too large and unwieldy, given that the Roman ships were at least a bowshot from the shore. A set of small, square mirrors, however, each held at exactly the right angle, would have behaved like one large mirror. If Archimedes had arranged a tight formation of soldiers on shore, each holding his own mirror, a beam of sunlight hot enough to ignite wood could have been created.

But how would each soldier know exactly where to aim his mirror? In 1973, a physicist named Albert Claus came up with a theory: maybe each soldier had a two-sided mirror with a tiny hole cut in the middle. If each soldier aimed the mirror so that the Roman ship was visible through the hole, the spot of light from the hole would fall on the soldier's cheek. The soldier would see that spot in the reflection on his side of the mirror. If he adjusted the mirror so that the spot disappeared into the hole, he would be aiming the light directly at the ship.

Now imagine hundreds of soldiers all doing the same thing. That would make for one seriously hot weapon—although a recent analysis suggests it would have taken 420 soldiers, each holding a mirror the size of a card table, to reflect enough spring sunlight to light a Roman ship. Or would it?

Many people have tried using mirrors to light things on fire. The most notable pyromaniac was the great scientist Comte

de Buffon, who in 1747 arranged eight-by-six-inch (twenty-by-fifteen centimeter) mirrors and focused them at a target of wood smeared with tar. On a clear spring day in Paris, 128 of these mirrors caused a piece of wood 150 feet (45 meters) away to burst into flames—smoke and mirrors indeed. Buffon tried other distances and numbers of mirrors and was successful at igniting a fantastic range of items.

Science Fact! *In 2013, the owner of a Jaguar in London, England, returned to his parked car to find that the side mirror and panels had been damaged; the car had partially buckled, and there was a smell of burning plastic. As it turned out, the glass on the side of a concave building nearby had focused the sun's rays on the car. The building has since been furnished with shades. The Jaguar has been repaired.*

In the early 1970s, a group of Greek sailors standing on a beach holding bronze-coated mirrors set a plywood silhouette of a Roman ship on fire from about 165 feet (50 meters) away. But more recent tests, including ones by MythBusters and MIT, produced less exciting igniting. The MIT team managed to set fire to a model ship after about ten minutes. MIT and MythBusters together generated lots of charring but very little flame.

While these experiments have shown that wood can be ignited by mirrors, the circumstances have to be precisely orchestrated to ensure success. And it's very unlikely that would have happened in the harbor at Syracuse. After all, the Roman ships would have been tossing and turning in the wind, making it almost impossible to focus a sunbeam on a particular spot long enough to cause ignition. Plus, the wood of the hulls would have been wet, and clouds would have dramatically diminished the power of the beam being reflected.

Given all of that, and given that Archimedes had already invented ingenious war machines—like a giant claw that could hook the prow of a Roman ship and tip it on end—it's unlikely that he would have spent a lot of time trying to make such a fallible, unpredictable technology as sun and mirrors work.

Also, it's not even clear that Archimedes ever had the mirror idea in the first place. The historian Polybius, who lived at the time of the battle, described a number of technologies that Archimedes invented and deployed against the Romans, but mirrors were not among them. Two other contemporaries who wrote about the siege also failed to mention mirrors.

A mathematician named Diocles wrote about "burning mirrors" in the decades after 213 BC, and while he did mention Archimedes in a mathematical context, he made absolutely no mention of mirrors in the siege of Syracuse. What's more, Diocles claimed he was the first to prove that a parabolic mirror could concentrate the sun's rays.

In the face of this tidal wave of uncertainty, one tiny refuge is that mirrors could have been used to momentarily startle and blind sailors. That would have given the Syracusans enough time to deploy some of Archimedes's other fine weapons of war.

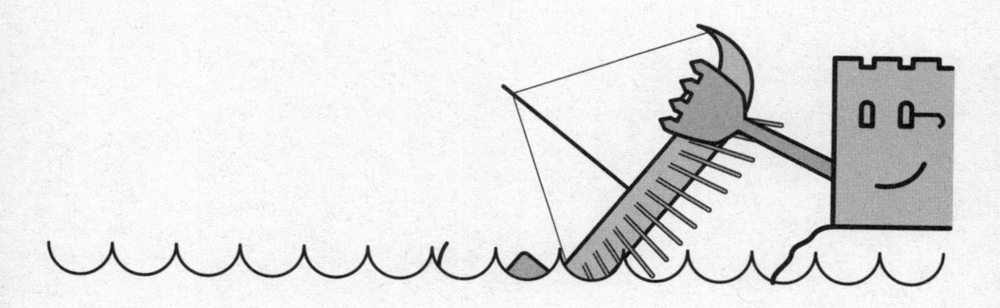

Science Fiction! *Science fiction author Arthur C. Clarke wrote a story called "A Slight Case of Sunstroke," set in the fictional country of Perivia in South America. Half the seats in a stadium are occupied by soldiers with souvenir programs silvered on the back. After a particularly bad call by the referee, the soldiers deploy their mirrored programs, focus the sun, and incinerate him on the spot. Not likely, but entertaining!*

Part 3
The Animal Kingdom

How do electric eels shock their prey?

The electric eel is an amazing creature on many different levels. Yes, it is basically a living battery, and it can use its electricity to detect, shock, and immobilize its prey or defend itself against predators. But it's even more bizarre than that: while it's a fish—big ones are up to six feet long—it has no scales and it breathes air. An electric eel must rise to the surface to breathe about every ten minutes or so.

Still, it's the battery that's the most interesting part. It's easy to ignore the fact that most living creatures, or at least those with more than one cell, run on electricity. Every single nerve impulse—and there are billions of them happening right now in your body—is a tiny jolt of electricity. The electric eel has taken this fundamental biological property and raised it to a high art.

The electric eel's body is not unlike a typical battery—say, the AA battery that powers so many household gadgets. Amazingly, about

four-fifths of the fish's body is filled with stacks of tissue composed of specialized electricity-generating cells. Even though each cell produces only about one-tenth of a volt, there are thousands of them, and the layering allows the voltage to build from one end of the eel's body to the other.

Like a battery, there is a positive end, at the eel's head, and a negative end, at its tail. Just as you have to close the circuit in, say, your flashlight by flicking a switch to connect the two, the eel accomplishes the same thing by allowing its electrical current to flow from its head to its tail through the water, which is a good conductor. Unlike a battery, though, when the electric eel releases current, it is sudden and powerful rather than steady and long-lived.

But not that powerful. Yes, the electricity from electric eels is high-voltage, but the resulting current is relatively weak and short-lived. It gives a human a painful shock but one that's very unlikely to be fatal. In fact, the eel is a little like a Taser. Tasers deliver about nineteen high-voltage pulses per second. Electric eels produce many more, about four hundred per second. The effect is the same, whether on a human or fish: paralysis, since the electricity causes muscles to contract spasmodically. The Taser has a greater impact, though, because the shock is delivered through two metal darts that actually penetrate the skin and deliver it directly, while

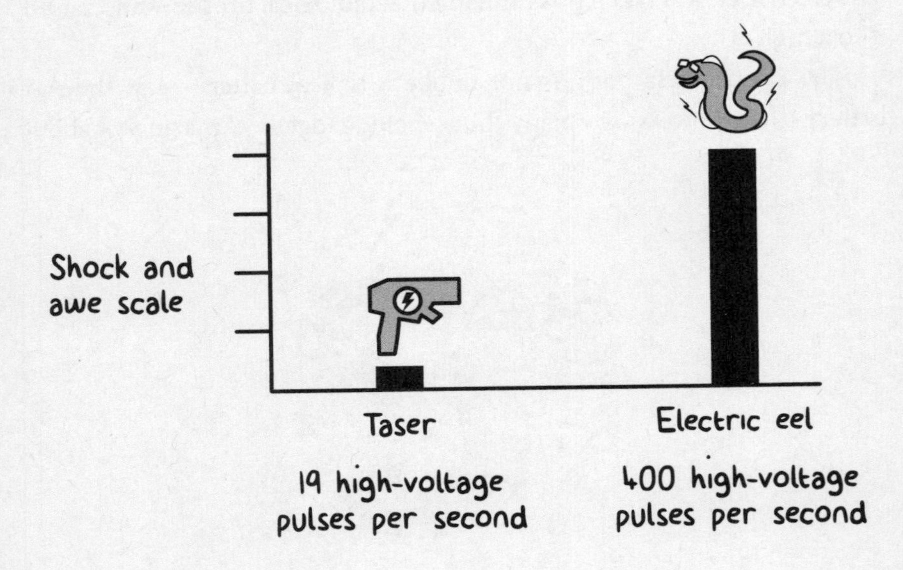

the eel has two disadvantages: it has to rely on its pulses traveling through the water, which offers resistance, and the pulse is emitted into the water without much targeting, so can't be anywhere near as focused—or concentrated—as the Taser.

That's not the end of the story. Ken Catania of Vanderbilt University has revealed that the electric eel hasn't just exploited electricity—it has refined it to an astonishing degree. It can turn the dial up or down, depending on exactly what it wants to do.

First, a low-power jolt allows the eel to locate its prey. When an eel emits one of these low-energy blasts, the muscles of a fish—even one that's hidden from view—involuntarily twitch. Eels are incredibly sensitive to any sort of disturbance in the water, so that twitch reveals the fish's presence.

Once the eel knows where a fish is, it can close in and power up the electricity and actually paralyze the fish; then it's game over. The eel's ability to temporarily paralyze a fish in this Taser-like way is obviously a pretty cool way to hunt and is extraordinarily effective: the blast only takes three milliseconds to take effect.

If this weren't enough, the electric eel also uses a clever technique to ramp up the delivery of electricity by curling its body in a C shape around any prey that is more difficult to subdue. The prey ends up trapped between the positive (head) and negative (tail) poles of the eel's body, with the electric current flowing directly between them both. This doubles the effective hit—already much higher than that delivered by a typical wall outlet—and most prey is unable to withstand that.

Did You Know . . . The great scientist and explorer Alexander von Humboldt traveled to the Amazon rain forest in 1800 to study electric eels. He was curious about how the eels hunted (electricity wasn't well understood at that time), and some local fishermen accommodated him by fishing for electric eels using horses.

According to Humboldt, the fishermen drove the horses into the shallow pools where the eels were hanging out. The eels' response was strange. Rather than fleeing, they shocked the horses, but not in the same way they would shock a fish. Instead, the eels reared up and actually made contact with the horses, delivering what looked like an extremely powerful shock, strong enough to fell some of the horses and drown them. After a while, though, the eels' electrical stores were exhausted and they could be collected by the fishermen.

One day, Ken Catania was using a metal net and protective rubber gloves to move his lab eels around. As he worked, he was amazed to see that the eels would sometimes leap out of the water, hit the net, and emit a series of incredibly high-voltage pulses. They probably feared the metal net was a predator: again, the eels use their electric discharges not only to kill prey but also to defend themselves against predators, and they likely determine the difference by the electrical conductivity of the unknown creature: small conductors are prey, large ones are predators.

When an electric eel sees a potential predator (like a large metal net), it can employ its standard approach and rely on water to conduct its electricity, but that might not provide enough current to deter an attack. By making physical contact with the attacker like a Taser, the eel's electricity would flow directly into the attacker's body, rather than through the water, making the shock substantially more powerful. And the higher up on the body the eel gets, the more electricity it appears to deliver, hence the leap out of the water, both in Catania's lab and Humboldt's pond. It would be a useful defense when creeks and streams become shallow in dry seasons, and prey or predators aren't fully submerged in the water.

Everything about the electric eel shows that the animal is not just an electric generator but a fine-tuned detection and killing machine. One thing that's still unclear, though, is why electric eels don't shock themselves. Maybe they're well grounded. At least, that's the current story.

Where do cats come from?

Cats, like dogs, have ancient origins, but the route they took to get where they are today was very different from that of their canine nemeses.

Ten-thousand-year-old cat bones were found buried alongside humans on the island of Cyprus. Those cats didn't swim there, so they must have been taken there by humans. And given their esteemed burial spot, they were clearly pets. It's assumed that domesticated cats date back to a few hundred, or even a thousand, years earlier than that, making their debut around eleven thousand years ago.

Science <u>Fiction</u>! *Although many believe the house cat comes from Egypt, the Middle East is more geographically accurate as the home of the first domesticated kitties. Egyptians definitely worshipped the feline, though, elevating them to god status. Cat mummies abound, and archaeological digs have even uncovered cat cemeteries.*

Genome analysis has identified the wildcat as the likely ancestor of the house cat—and one species in particular. The domestic cat lines up genetically with the Middle Eastern version of the wildcat, considered the least fearful and most gentle of all of its kind, and therefore the most likely to be domesticated.

There is another factor that leads us to conclude the Middle East is the likely geographical origin of the house cat—the rise of agriculture. Ten thousand years ago or so, our ancestors abandoned nomadic life and began to settle on the land, growing and storing their food. The downside of storing grain is that you are forced to share it with rats and mice. Small rodents are the perfect diet for wildcats, and so, at least the theory goes, granaries attracted mice, which in turn attracted cats to agricultural settlements. Perhaps cats and humans didn't interact much at first, but there would have to have been some accommodation on both sides, and why not? There were obvious benefits to both. It's also quite likely that at some point, kittens from the granary were adopted as house pets. These lucky cats probably ate scraps from the dining table and forged closer bonds with people.

Did You Know . . . Black cats are everywhere at Halloween, for good reason. In the Middle Ages, the Catholic Church sought to eliminate heretical pagan groups such as the Knights Templar. To pagans, cats represented fertility, but Christians associated cat-worship with the devil, cannibalism, orgies, and the execution of children. This association lasted for centuries, and it's a testament to the species that cats survived, despite the incredible persecution they endured.

Examination of the cat genome reveals the changes wrought since felines moved in with us, although these adaptations are not as widespread as those in dogs (because cats are more recently domesticated). Cats continue to retain many of their wild attributes: the broadest hearing range of any carnivore (into the ultrasonic to hear those little mouse conversations), fantastic night vision, and adaptation to a fully carnivorous diet. However, the cat genome shows clear

changes in areas of the brain that regulate timidity, fear, and reward-seeking. Overcoming timidity and learning to beg for rewards would have been helpful in ensuring their place with us.

Did You Know . . . Cats have lost the supreme sense of smell that dogs have, which is understandable because they are primarily visual hunters. At the same time, they have amped up their nasal detection of pheromones, the subtle chemical signals that regulate social behavior.

Having originated in the Middle East, cats from that founding population were taken around the world by humans. The riot of coat colors that domestic cats now exhibit was bred into them by us over the last two centuries—a blink of an eye in the evolution of an animal. Is it possible that over time cats will become even more domesticated? Who knows? Perhaps one day, they'll all roll over onto their backs and let us rub their bellies the way dogs do. Or maybe not.

Where do dogs come from?

Dogs came from wolves. That seems straightforward, but if you look at a Chihuahua, a Great Dane, and a wolf side by side, you'd be right to wonder if there's more to the story of dogs' ancestry. Knowing that dogs came from wolves is just the start; it's why, how, and when that happened that are the really interesting questions.

Did You Know . . . There are at least 80 million dogs in Canada and the United States alone. There are roughly 6 to 7 million dogs in Canada, and another 75 to 80 million in the United States.

The first attempts to figure out where dogs came from relied on archaeology. Researchers have uncovered skulls of doglike animals that date back to more than thirty thousand years ago. The skulls they found were a long way from anything we'd recognize today, though. At least in the very early transition from wolf to dog, several physical features of the animal head's changed: the teeth became more crowded and the skull grew wider. To put this into perspective, if modern dogs were on the scene more than thirty

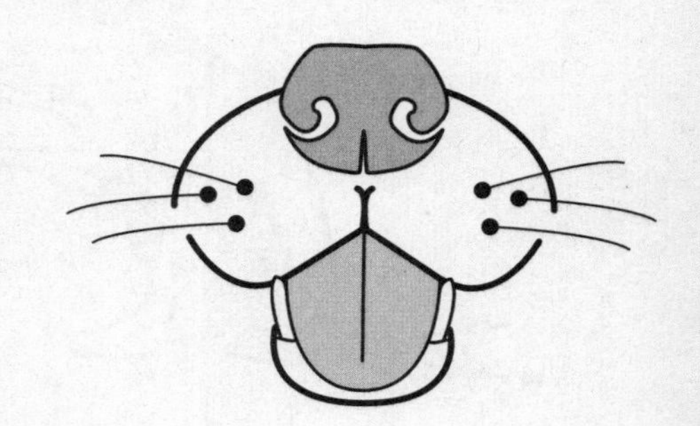

thousand years ago, that means they were in Europe shortly after the Neanderthals died out. That great depth of time sheds some light on how the wolves of the time might have become domesticated.

There used to be a popular idea that one day, an enterprising ancestor of ours came upon a den of newborn wolves, scooped one up, and tamed it, giving rise to the millions of dogs in the world today. But the reality is that wolves played as much of a role in their taming as we did.

Thirty thousand years ago, our ancestors were still nomadic hunter-gatherers, constantly relocating as they searched for prey such as deer, smaller mammals, and edible plants. The wolves of that time—and most researchers now agree that they were not the same animal as the modern wolf *Canis lupus*—were also on the move, looking for prey that they would attack and devour as a pack. Packs of wolves and humans on the hunt would undoubtedly have come into contact. Humans wouldn't have looked on wolves as prey and nor would the wolves have seen humans that way. But their proximity means that wolves would have encountered carcasses left behind by humans. Smart animals are able to put two and two together: humans (inadvertently) provide food. So, the theory goes, packs of wolves would have started to drift along with human hunting parties to pick up their leftovers. And the humans would have benefitted from keeping the wolves close—packs of the animals living on the edge of their hunting camps would have alerted the humans to the presence of either predators or prey nearby.

Did You Know . . . There was a huge population crash among the wolves that gave rise to dogs. It happened about fifteen thousand years ago—the same time as the conclusion of the last ice age—as a result of an unknown environmental incident. The crash might explain why no one has found any remains of the wolf ancestor of modern dogs.

These archaeological theories were useful up to a point, but they didn't explain exactly how the wild wolves of the past turned into the dogs we know today. To solve that problem, scientists had to turn to genetics. One of the earliest experiments to use this new approach was run in the late 1950s by Dimitri Belyaev in the Soviet Union. Belyaev believed that the tamer an animal was—or the more easily it got along with humans—the easier it would be to domesticate. To test his idea, he began a program of breeding silver foxes based on how tame they were. He started the program in the 1950s and worked on it until he died twenty-six years later. By evaluating how eager a young fox was to accept a human's approach or how willing it was to hang out with a human and not bite them, Belyaev could establish how tame that animal was. After examining each generation of foxes, Belyaev would allow only the top 20 percent of them—the ones he found to be the least fearful and aggressive—to breed.

As the years went by, the foxes grew increasingly tamer with each generation, but they also changed physically and behaviorally: they started whimpering to attract human attention, licking their trainer's hands, and wagging their tails. Their heads changed shape, their ears became floppy instead of standing up straight, their tails started to get curly, and their fur changed color. Even their hormones changed from those of their wilder ancestors.

Belyaev's experiments suggested to scientists that a similar kind of selection might have happened around our ancestors' campfires. The less aggressive and less fearful wolves would have established themselves as part of the human camp and bred with one another, making each new generation tamer and changing their physical features until they became the dogs we know today. Recent work by scientists hasn't been able to determine exactly when this change happened. One study found that wolves finished their evolution into the modern dog around eleven thousand to sixteen thousand years ago. Another study sequenced the genomes of fifty wolves and dogs from around the world and concluded that the domestic dog originated in Southeast Asia around thirty-three thousand years ago.

Science Fiction! *Some theories have claimed that dogs didn't appear until humans began engaging in agriculture. Scientists used to think that dogs' genetic code had a starch-processing enzyme, amylase, which wolves lacked. Because starch would be found only in food or waste left over from farmers, not hunter-gatherers, the scientists reasoned that dogs could only have developed after humans gave up their hunter-gatherer ways to take up farming. That idea was put to rest when recent genetic tests found that wolves do indeed have the amylase gene.*

In the course of dogs' evolution, they developed one of their most distinguishing characteristics: their bark. Fully grown wolves hardly ever bark, but wolf pups do. Somehow, as dogs became domesticated, they retained that juvenile characteristic as a part of their adult lives. Not only that, but dogs refined barking into a useful communication tool.

Sophia Yin and Brenda McGowan at the University of California recorded a grand total of 4,672 barks uttered by a mix of ten breeds. They recorded the dogs' barks in three different situations: when a stranger rang the doorbell, when a dog was separated from its owner by a locked door, and when the dog was playing with either a human or another dog.

When they analyzed the barks acoustically, they found that there was a wide variety ranging from harsh low-frequency ones to higher-frequency barks rich in harmonics. The barks directed at the stranger ringing the doorbell were harsh and long in duration, but the barks while the dogs were separated or playing were higher frequency with long spaces between them. There is a general rule in the animal world that a harsh sound means hostility, while higher-pitched, musical sounds are more social. By that measure, dogs appear to be wilder than we thought.

Wherever and whenever dogs originated, it's obvious that they eventually migrated all over the world, either on their own or with humans. But although it's been a very long time since the wolf and the dog separated, the two species can mate and have fertile offspring even now, so genetically they are still extremely similar. So similar, in fact, that their mitochondrial DNA—the genes that are passed down strictly through the maternal line—is 99.9 percent the same. Half of that 0.1 percent difference, though, is linked to the brain and social interactions. So you can thank that tiny fraction the next time Fido licks your hand instead of biting it.

What attracts mosquitoes to me? (And what can I do about it?)

Mosquitoes are more annoying than dangerous for most of us in North America. I don't mean to trivialize West Nile virus, nor am I ignoring the fact that with climate change, more mosquito-borne diseases are beginning to creep into the southern United States. But for most of us, it's those buggy summer nights that make us wish them extinct. That's not likely to happen any time soon, so in the meantime we search for ways to make ourselves unattractive or, even better, invisible to our biting nemeses. There's just one problem: mosquitoes are finely tuned for searching us out—so much so that after reading this, you might lose faith that you will ever find peace without taking refuge indoors. And even then . . .

Mosquitoes have a package of detectors and responses that would challenge the most powerful computers linked to the most sophisticated robots, all operated by a brain that is about the size of the period at the end of this sentence. These detectors include human-targeting sensors that are turned on in sequence as the mosquito closes in on a person.

Imagine you're sitting outside at dinnertime in the summer in an area frequented by mosquitoes. Assume there's not too much wind, no campfire—just you. You are breathing—that is, exhaling carbon dioxide. As that plume of gas wafts through the evening air and breaks up, a wisp of it passes over a mosquito's head, coming into contact with specialized CO_2 receptors on her maxillary palps (antenna-like extensions on her head). A volley of nerve impulses is triggered in the bug's brain. She's on your trail.

Now she moves toward the source of the carbon dioxide—you. Not only has she detected the gas, she's adjusting her overall flight direction toward higher concentrations—all of this even though carbon dioxide sensors on the palps represent only 2 percent of her total sensory array. But it's not enough for her to depend on finding you by CO_2 alone—otherwise she would fly directly into your mouth, and while that would be annoying to you, it wouldn't help her at all. At some point she needs to switch her attention to other attractants to be able to close in on your bare skin.

Did You Know . . . Only the female mosquito bites. She needs the protein in your blood to develop her eggs.

In a beautiful set of experiments, Caltech researchers showed that once a mosquito detects carbon dioxide, she then pays attention to visual features in her environment. To prove this, researchers had female mosquitoes fly freely in a wind tunnel that was visually plain except for a single dark square on the floor at the end (mosquitoes' visual systems are more sensitive to dark than light). In the presence of normal air, mosquitoes paid no attention to the floor at all, but when the

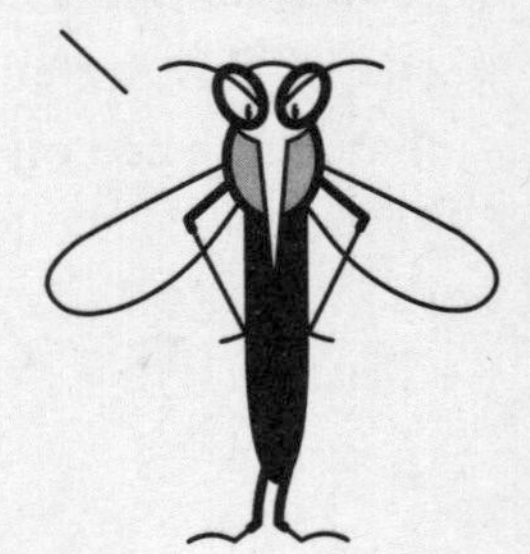

experimenters introduced carbon dioxide into the tunnel, the mosquitoes immediately began to inspect the dark square on the floor, spending hours hovering about an inch above it. They had switched over from gas detection to vision, trying to home in on a potential target.

Science Fact! *It's true that wearing light-colored, high-contrast clothing will help you stay invisible to mosquitoes, but it messes up only one of their many detection systems. Scientists agree that wearing light clothing alone is not enough to thwart the "annoyingly robust" sensory powers of the female mosquito.*

In the Caltech experiments, once the mosquitoes closed in on a target using vision, they then deployed heat- and chemical-sensing to find the right landing spot. At that stage, warm objects attracted mosquitoes much more strongly than ones at room temperature. Finally, mosquitoes were attracted even more strongly to warm things with high humidity—like your sweaty skin.

And just in case you're thinking there might be a way of circumventing this string of sensory modules that turns you into guaranteed mosquito bait, think again. As it turns out, a mosquito can even skip one of her detection tools and still find you without much trouble. Let's say you're holding your breath so she doesn't detect your CO_2; she can still be attracted to your body heat as long as she flies close enough to you. And all she really needs is someone close to you to exhale and she'll be on her way over to you and your breathing buddy. If you could persuade your friends to stop breathing alongside you, that might help you stay incognito—but not entirely. If you're warm and sweaty and happen to be wearing dark clothing, you're still giving yourself away.

Why do lizards shed their tails?

Of all the tactics you might use to defend yourself against attacks (run away, fight back, play dead), surely one of the strangest would be to simply rid yourself of a body part! Lizards are pretty fantastic at this. It's called autotomy—the defensive strategy of letting the tail go if it's seized by a predator.

When a snake attacks a lizard from behind, it bites at the tail right where it joins the lizard's body. But before the snake can swallow its prey whole, the rest of the lizard is gone, having fled while the snake was distracted by the wildly thrashing tail in its mouth. The lizard lives another day!

But as always, freedom has a price. Without a tail, a lizard's mobility is compromised, especially when climbing. Also, its mating display might be less effective. And in a temperate climate, losing a tail late in the season might mean a crippling loss of the animal's fat stores—stores that are essential to get the reptile through a winter of hibernation. The tail can be so crucial in this regard that some lizards take the chance of being attacked again and return to consume what's left of their own tail after a predator has abandoned it. Also, there's the problem of a second attack. It will be a sad tale next time when there's no tail for the lizard to drop.

The good news? Lizard tails regenerate. But that takes a lot of time and resources. So that's the balance that is struck: lose a tail to avoid almost certain death, only to raise the risk of dying months later.

Did You Know . . . Some lizard species, such as the five-lined skink, have bright blue tails when they're young that later turn boring shades of brown. Why? It's possible that as the animal matures, camouflage becomes the preferred method of avoiding predators rather than tempting them with a bright blue morsel.

If a predator bit down on your arm and you ran away leaving your limb in its mouth, you can be sure you'd be in pretty dire shape. So what's different for a lizard? A closer look reveals how the unique anatomy of the lizard tail allows for autotomy. The tail is precut, divided into a set of fracture points. There really isn't a lot holding the tail together; even the muscles are fitted loosely, not firmly attached to each other. When the muscles around any of these preset fracture points contract, the vertebra breaks and the whole organ falls off cleanly, without leaving a bloody pulp behind.

Of course, escape is wonderful, but the really fantastic part of autotomy is what the tail does after it separates: the tip keeps flicking back and forth at pretty amazing speeds, and every once in a while, at least in some species, it suddenly lashes out. It is alive! Well, no. It isn't. But it does have self-contained neural circuitry and enough stored energy to keep moving for a minute or sometimes even more—long enough for the rest of the lizard, the living part, to run away.

It's curious that the frantic movements of the detached tail have different effects depending on the predator. Experiments with domestic cats showed that the vigorous movements of the detached tail confused or at least occupied the cat, allowing the tailless lizard to escape. With snakes, tail lashing prompted concentration on subduing the tail's movements, thereby allowing the rest of the lizard to escape.

In the end it's not a bad bargain for predator or prey: the predator gets a nice fatty piece of lizard; the lizard gains freedom—at least for a while.

Why do some creatures throw their feces?

There's a caterpillar that does it brilliantly, a penguin that is pretty good, and an ape that is truthfully more interested in the act of throwing than in what's being thrown. I leave it to you to decide which creature is more accomplished at shooting the shit.

That caterpillars, at least some of them, can shoot their own shit like a projectile weapon likely comes as a surprise, but a closer look (not too close!) will convince you that (a) they do it in a sophisticated way and (b) it is essential for survival.

The skipper caterpillar is the best insect shooter. It's a leaf roller—that is, when it settles on a leaf to feed on, it pulls one side of the leaf over

itself, just as you pull the covers over you when you get into bed. The only difference is that you don't anchor the covers in place with strands of silk as the caterpillar does. Once sequestered this way, the caterpillar is invisible, and predators that search for it using sight will not find it. But smell is another thing, and what gives the caterpillars away is the odor from their frass, which is a nice entomological word for excrement.

Some cool experiments in the lab have shown that wasps interested in a caterpillar meal will spend far more time inspecting rolled-up leaves containing frass than those without frass. So it pays the caterpillar to ensure that its frass is deposited far from its body. It's safety, not cleanliness, that has promoted this skill (and has earned this insect the nickname "scatterpillar").

Researchers more than a hundred years ago noticed that skipper caterpillars launch their poop at a decent velocity several body lengths away, and in a variety of directions, too, so they don't end up with a frass pile, making it obvious that there's a caterpillar in the vicinity. It took until the 1990s to clarify the actual mechanism by which they literally shoot the shit. The caterpillar's rear end is fitted with a complex of plates, "combs," and pressure vessels, all of which are synchronized to be able to send the shit flying.

Imagine, if you can, a fecal pellet ready for expulsion. It is maneuvered into position by a combination of fleshy collars on the creature's back end. These move the pellet into contact with the anal plate and the anal

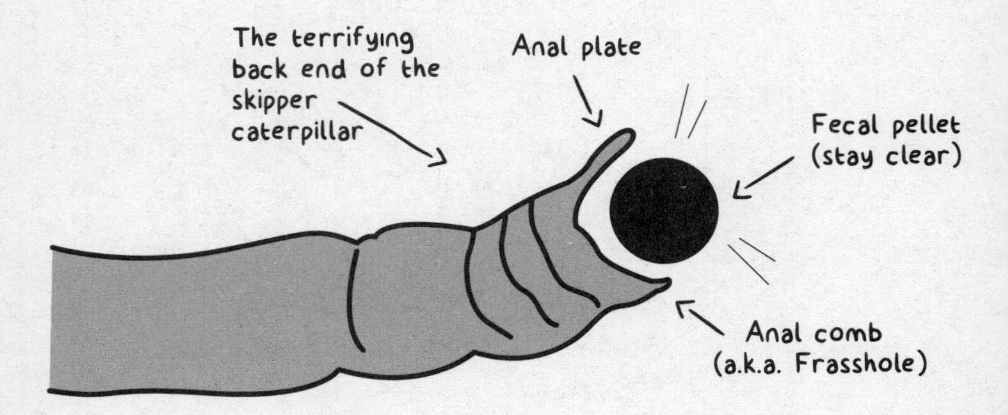

comb. The plate is the launchpad and the comb is the trigger, the crucial piece, hooked as it is on one of the collars until, with a rapid buildup of blood pressure (actually not blood but the insect version, hemolymph), the comb suddenly slips off its mooring and the plate lunges forward, sending the frass flying. It might sustain launch forces of more than one hundred g's as it exits the caterpillar's body, at speeds in excess of six feet (two meters) per second. The mechanism has been charmingly described as similar to the game of tiddlywinks, where hard pressure on an object sends it flying.

It's a big jump, evolutionarily speaking, from a skipper caterpillar to the Adélie penguin, but that's pretty much how far you have to go to find another creature with at least a modicum of skill in this fecal volley game. Adélie penguins (and chinstrap penguins) eject their feces from the nest—a feat they accomplish simply by perching on the edge of the nest, rear end outward, bending forward, lifting the tail, and letting it fly. But sadly, the forces and distances, at least with respect to body size, aren't nearly as impressive as the scatterpillar's.

Of course, not all poop is created equal, and in this case, the bird's waste is not solid but liquid, with the viscosity of something like olive oil. A typical nest-side shoot might be a little over a foot (about thirty centimeters), and unlike the compact little packages ejected by the caterpillars, this is more like a streak of sluice, colored pink if the last meal was krill or white if it was fish. It isn't as clear why penguins volley their feces, but the best explanation is that the birds spend a lot of time preening and cleaning, and getting rid of feces would be consistent with that. Incidentally, the streaks left behind radiate from the nest in all directions, but it's not clear whether the penguins do this on purpose.

And now, an evolutionary side step to one of the biggest land mammals: the hippopotamus. You don't have to venture far onto YouTube before you find videos documenting the delightful hippo habit of vigorously wagging the tail while defecating, ensuring that its feces fly off in all directions. I've seen this myself, and I can tell you it's impressive.

The mechanism isn't particularly sophisticated: if you were to fire blobs of mashed potatoes at a rapidly swinging pendulum (and I mean rapid), you'd get pretty much the same result. But why would you do that? And why does the hippo do it?

Most explanations are guesses. Spreading feces around might be a display of dominance by high-ranking animals; it isn't likely to be territorial, though, like dogs and hydrants, because male hippos' territories are in the water—a stretch of river, for instance. There'd be no point tail-swishing in the water because the feces would be taken away by the current anyway. Could it be just for amusement?

Chimpanzees love a good fecal throw—in the wild and in zoos. Zoo-goers know that there's always a good chance that chimps will throw their poop around, to the horror and/or delight of spectators. But they don't restrict themselves to poo, and their throwing skills say more about their brains than they do about what they're throwing. For ten years, researchers watched a chimp named Santino at the Furuvik Zoo in Sweden. Santino liked to express his general displeasure by throwing rocks, not poo, at visitors. But it was what he did when he wasn't throwing that intrigued the investigating scientists. When the zoo was closed, Santino would scout his territory for rocks that he could pile up in anticipation of new visitors arriving the next morning. He'd hoard these rocks, throw them when the visitors arrived, then stockpile at night once more—over and over in a cycle. Researchers drew the conclusion that his future-planning had a human-like quality.

This story intrigued me because I, too, was once the target of rocks thrown by an angry male chimp. I was in Japan, at the Primate Research Institute of Kyoto University, shooting for *Daily Planet*. The institute's director at the time was Tetsuro Matsuzawa, known worldwide for his work with both captive and wild chimpanzees. He showed me and my team a fantastic chimp named Ai that easily outscored me on a memory test. While Ai's memory scores were impressive, I also met an alpha male who had a pretty good underhanded throw.

As we approached, the alpha male retreated to the far side of his enclosure, looking anxious. Then he started shaking, picked up some stones, ran to his left, and started hurling them at us. Matsuzawa thought it was hysterically funny. I was concentrating on not getting hit!

Whether it's rocks or feces, it's the throwing that interests scientists. In captivity, chimps seem to do it to vent, or for the entertainment value. It's the only time they can really exert an influence over visitors. In the wild, chimps throw at each other, but not usually with enough power and accuracy to injure. It's really about sending a message—not the most friendly one, either.

One recent study showed that the brains of chimps that throw well are visibly different, at least as seen through the eyes of an MRI. They have more elaborate sets of connections and faster, more efficient signaling on the side of the brain that controls the throwing arm, which is most often the left-brain-to-right-arm connection. And most interesting, the good throwers were also the better communicators, more adept at tasks that demand the chimp be aware of the researcher and communicate with that person.

In human beings, that cross-body control would also most often be a left-brain-to-right-arm connection. But we also have language, and there's an argument that the fine motor control of the lips and tongue for speech is a refinement of the neural circuits that control throwing. Maybe that's the reason the left side of the brain controls speech: it co-opted the already elaborate motor control systems on that side.

So throwing first, then speech (as opposed to language, which doesn't have to be spoken). Of course, it's also easy to imagine that gestures could provide the critical bridge between throwing and speaking. It's not the splatter of feces on your coat that is significant; it is the fact that the chimp is giving you a glimpse of how the human brain evolved. This is what sets the chimps apart from all the rest: yes, feces are thrown, but with intent.

Why do geese fly in V formation?

The most popular theory for why geese fly in a V formation has always been that in a such a formation, every goose gets a slight boost from the others—a literal updraft, a vortex of air that swirls off the tips of the wings of the birds next door.

If efficiency is the goal, why wouldn't geese instead fly lined up like a chorus line, each bird directly beside the other? The aerodynamics equations suggest that as few as nine birds flying in a chorus-line formation would gain 50 percent more range. But a V formation distributes the energy savings among all the birds a little more fairly: the goose in

the lead benefits a little less (although still gets a bit of a kick), and birds farther back in the V fare better. That might be what makes it the go-to goose formation.

Some experts have suggested there could also be a social reason for the V: it's a perfect formation for keeping tabs of one's wingmen. But considering there are many ways for geese to maintain social cohesion, it still comes down to the V giving all the birds a lift.

A V formation is a robust thing, too. If a goose flies too far forward out of the V, she'll lose uplift and naturally fall back into line. If she drifts too far back, she'll get more lift and suddenly find herself surging forward into her proper place.

Did You Know . . . Fighter pilots experience a similar boost to geese by flying in a V formation. They save as much as 18 percent in fuel consumption by flying just behind the wingtips of the plane in front of them.

The positioning is crucial, because besides the much-coveted lifting vortex at the wingtips, there's also a "downwash" spilling off the back of the wing that would make it harder to fly. That's a good place to avoid. To work best, the V should have an angle of about 100 degrees—about the angle your little finger makes with your thumb if you stretch your hand as wide as possible.

For a long time, that's about all anyone knew: that the V was an energy-saver. But in the 1970s and 1980s a small team of dedicated naturalists decided to analyze the spacing of geese as well. They spent their days standing out in the cold, cold autumn, desperately hoping for a V of geese to pass directly overhead so they could photograph it and measure the angles and distances of the wings. Getting geese to fly where you want them to, when you want them to, is not exactly easy. This team got photographs all right, but interpreting the distances in the photos was challenging and contentious.

While they were figuring that out, new theories sprung up. One posited that the angle of the V was less important than the relative positions of the wingtips. Each bird's wingtips should be directly behind the wingtips of the bird in front. Researchers went back to their photos to check out this possibility and concluded that, yes, that's exactly how the birds were positioned. So maybe the secret was wingtip arrangement, not V angle. But there were still puzzling observations of geese flying in ragged Vs, or Vs with more birds on one side than the other. The question remained: Why?

The breakthrough came when two unrelated groups—European scientists interested in the aerodynamics of flight, and conservationists working with a rare bird called the northern bald ibis—found each other.

This ibis disappeared from the wild in Europe in the 1600s, and a group in Austria was trying to reintroduce them by hand-raising young birds and then guiding them along their historic migration route with an ultralight aircraft. This is almost exactly what Bill Lishman, also known as *Father Goose*, did in the 1980s and 1990s, when he trained geese, and later whooping cranes, to follow ultralights on a migration route. (His exploits were later dramatized in the movie *Fly Away Home*.)

Back in Europe, the new partnership between conservationists and scientists led to a great opportunity. Previously, the scientists had had a problem: the instrumentation they attached to the birds, which recorded position and wingbeat, worked only if the scientists could control where the birds landed. But how do you tell a flock of ibises where to land? Enter the conservationists. Because they led the flocks, they were able to guide the young ibises to land where the scientists wanted so that data could be downloaded at every landing spot.

And that data was gold. It showed that the ibises maintained exact spacing between each other in a V. But there was more than that: they also timed their wingbeats to take maximum advantage of the uplift provided by the bird in front. One of the researchers likened their efficiency to walking through the snow by placing your feet in the footprints of the person in front of you. There's a rhythm that maintains maximum lift, and these birds adopted it seamlessly.

But even more surprising was their adaptability when the formation shifted. These ibises never maintained a rigid V. Sometimes they even fell into single file. That should have been a bad thing, because they'd encounter the downwash off the wings of the bird in front, but they adjusted by automatically adopting a mirror-image wingbeat, in which up and down strokes were reversed. This limits the disadvantage of the downwash.

Interestingly, these birds were hand-raised by humans, so no adult ibises had taught them how to fly in a V. Still, they did it. It might be that they simply adjusted their positioning and wing strokes "on the fly" because it saved them energy.

So if flying in a V saves so much energy, why don't all birds do it? The aerodynamic theory suggests that the advantages accrue to large birds whose wings throw off big vortices. Sparrows, not so much. Smaller birds would have to fly extremely close to each other to get any uplift, and as far as anyone knows, they don't bother.

How do octopuses camouflage themselves?

You can have chameleons—when it comes to camouflage, I'll take octopuses, squids, and cuttlefish any day. Collectively known as cephalopods, all three can disguise themselves by changing their appearance and fading into their background.

Imagine how tricky that is: a predator is standing across from you, and your only escape is to replicate your surroundings so the predator's eyes slide past. I'll admit there is some pretty amazing advanced camouflage gear for hunters and soldiers, but if you're wearing it, you'd better hope you're in front of camo wallpaper so you blend right in. Cephalopods, however, move past multiple backgrounds from moment to moment, and

predators approach them from all directions, so they have to be able to adopt their pattern and color in a split second or they're lunch.

Did You Know . . . Cephalopods are color-blind. Their eyes are able to respond only to a very narrow wavelength in the blue-green part of the spectrum.

So how do they escape detection? Because they have brilliant biological tools at their disposal. Their skin amounts to a high-definition, pixelated, multicolored electrical sheet, with three layers of specialized cells. One layer is made up of colored cells called chromatophores, which are controlled directly by the creature's brain. Each has muscles that contract and pull, expanding its surface area by as much as five hundred times, dramatically increasing the amount of color the skin displays. Considering there are millions and millions of chromatophores in a cephalopod's skin—at least 10 million in a cuttlefish—this is a potent system.

Below the chromatophores are two layers of reflector cells, both of which are colorless. The first layer reflects light in the same way a soap bubble does. You might have noticed that as a soap bubble drifts through the air, it changes color, from red through orange to blue and eventually, just before it pops, black. Those changes are the result of the soap film getting thinner and thinner and changing the wavelength of the light it reflects. Cephalopod skin has cells with stacks of plates in them; the thickness of the plates, together with the width of the spaces between them, creates soap bubble–like interference with light, which generates an array of colors. The reflector cells in the second layer are brilliantly reflective spheres, like tiny disco balls, that reflect light of all wavelengths from all directions.

By adjusting the size of its chromatophores, a cephalopod can produce three patterns: uniform, mottled, and disruptive. The uniform pattern is very fine grained and is produced by shrinking the chromatophores. An animal might use this as camouflage against sand. The mottled pattern is coarser, like a gravelly sea bed. The disruptive pattern is a dramatic display with big chunks of different colors: think of a seabed dotted with corals and rocks.

So how do cephalopods choose which of the three display variations to use? Experiments with cuttlefish show that sometimes they match the background on which they're resting. Other times they "masquerade" as a rock or other feature somewhere nearby.

Sometimes by doing that their outline seems to change and they don't look like a cuttlefish anymore. The challenge for any of these camouflage-dependent animals is to take into account the kind of predator it will most likely encounter. This has been called the "point of view" predicament: the predator sees a different scene from the prey. For instance, the octopus has to use the visual information it gathers from where it is to create camouflage that will work for both a fish attacking from above and for a moray eel attacking from the side.

A different kind of complication arises for free swimmers, like squids and cuttlefish. If a predator approaches from below, the cephalopods are backlit by sunlight; but even though their bodies are partially transparent, their internal organs aren't. In this situation, the reflector cells come into play, channeling light through their bodies, making their insides lighter and harder to see.

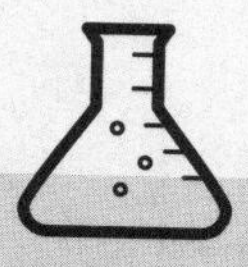

Science Fact! *All cephalopods are intelligent (although it seems octopuses might be the smartest of the lot).*

And what about that color-blindness? The colors produced by chromatophores are mainly yellow, orange, and dark brown (and their combinations), which are a good match for the environments these animals have colonized. So it's not so much that cephalopods see those colors and adjust their chromatophores to match but rather that they sense brightness and shading and depend on chromatophores that are already tuned to the approximate colors.

There's also some evidence that texture (fine grained or coarse) rather than color provides the best protection. In fact, some cuttlefish are able to perform camo magic in near darkness.

Cephalopods also use camouflage as a means of communication. What could be more effective for sending messages than sets of flashing colors? Besides, those reflector cells also polarize light, and most fish—among them these animals' predators—can't perceive polarized light, but cephalopods can. Not just a visible communication but a covert one as well!

Did You Know . . . Squids have such exquisite control over their displays that they can have a polka-dot body, a striped fin, and dark tentacles all at the same time.

Predictably, the military is interested in the mechanics of cephalopod camouflage and hopes to one day apply them to soldiers' uniforms. And understanding the cephalopods' ability to change color and pattern could even lead to new cosmetics and product lines one day, too.

How can a mongoose survive a cobra's bite?

In Rudyard Kipling's *Jungle Book*, the mongoose Rikki-tikki-tavi realizes, "If I don't break [the cobra's] back at the first jump . . . he can still fight." He looks at the thickness of the cobra's neck below the hood and knows that's too much for him and that a bite near the tail would only make the cobra savage. "It must be the head . . . the head above the hood. And, when I am once there, I must not let go."

Rikki-tikki's planning a surprise attack, but even so, it's risky. The venom of a king cobra, for example, can kill a person in half an hour. In fact, the amount in one bite from a king cobra can kill twenty people: it can deliver about a quarter of a shot glass of venom in one strike, more than any other snake. Happily for us, cobras prefer to avoid contact with people; their most common human victims are snake charmers.

And while mongooses are notorious tough guys—sturdily built, low to the ground, hard predators—Rikki-tikki is confident for good reason: he has biochemistry on his side.

The secret lies in the venom and how it works. A king cobra's venom looks like it was assembled by a fussy yet nasty mixologist: it's a combination of many different types of venom that target different organs, all in one convenient dose. One of those venoms is common to many species of snakes: alpha neurotoxin. It exerts its deadly effects where nerves meet muscles, but at an ultra-microscopic level, where one molecule comes into contact with another.

Nerves signal muscles to contract. There's a tiny gap between them, and when an impulse sweeps along the nerve and reaches the end, the nerve releases millions of molecules called neurotransmitters. These drift across the gap—a mere millionth of an inch—and plug into special receptors sitting on the surface of the muscle. Then the muscle contracts.

Once that happens, the transmitter molecules have to be cleared away to allow the process to be repeated. There's another molecule whose job it is to do that.

If this sequence—nerve impulse, transmitter release, muscle contraction, removal of transmitters—weren't happening all over your body all the time, you couldn't breathe, much less move.

Alpha neurotoxin targets this usually smooth-running system. Like the neurotransmitters, it's shaped to fit into the receptors on the muscle cell; unlike the transmitters, it doesn't get removed. That muscle cell will never contract again. As the venom spreads through the body, more and more contact points between nerves and muscles are blocked. It's said that even an elephant can be killed by the typical dose of king cobra venom.

So what does Rikki-tikki-tavi have going for him? Evolution. Over long periods of time, and with a great debt owed to many dead mongooses, changes to the mongoose's receptor molecules have made them resistant to venom. The venom molecules simply can't bind to the receptors the way they do with other animals, but the mongoose's neurotransmitters still can. So while the mongoose's speed and thick fur are both protective, once bitten, it has another powerful mechanism for self-defense. And you don't have to look far to see other, similar examples of creatures that have developed immunity. The honey badger has a good reason for its attitude. ("Honey badger don't give a s—t," it's been said.) The honey badger also has altered receptors, and that makes it a snake eater rather than snake prey.

Did You Know . . . Ironically, the king cobra is resistant to snake venom. Why? Because snakes are the cobra's primary prey and it has to be protected against them.

The final weapon in this array of molecular blows and counterblows is antivenom. To make antivenom, small amounts of snake venom are injected into domestic animals, like horses or sheep, and the animals' immune systems react by manufacturing antibodies. Those antibodies are then collected from their blood and concentrated. Once injected into a snakebite victim, the antivenom and venom molecules meet in the bloodstream and enter into what becomes a fatal embrace for the venom. It can't escape the antibody, making it impossible for the venom to attach to the muscle receptors, which means life goes on.

The mongoose's story is a classic example of evolutionary moves and countermoves. Right now the mongoose holds the upper hand in the evolutionary battle, but the fact that the cobra's venom contains so many different molecules with wildly different micro-architectures leaves the door open for one of those molecules to mutate into a deadlier form. When that happens, the tide will begin to turn.

Can an elephant jump?

There's a myth that elephants are the only animals with four knees. If that was the case, elephants might be great jumpers. But they don't have four knees: the bones in their front legs and our arms, for example, match one to one. They're the same kind of bones. So the question of whether elephants can jump has nothing to do with the number of knees they have; it's a question of force and pressure.

Huge animals like elephants, hippos, and rhinos have legs that are thick like columns, much thicker relative to their bodies than, say, the legs of a cat. The bigger an animal is, the more its proportions change. An elephant is about thirteen times taller than a cat but weighs eight hundred

times more. If it were the same proportions as a cat, that massive weight would snap its skinny little legs in two.

So how does the elephant support its enormous weight? With thicker legs. A cat's thigh bone is around 0.4 inches (about 1 centimeter) in diameter, but the elephant's is ten times wider (4 inches, or 10.2 centimeters). A larger diameter translates mathematically into a much bigger surface area, meaning that the weight-bearing surface of the elephant's leg bones is about twenty-five times that of the cat's. It's enough that the elephant doesn't break its legs, but the weight still puts a lot of stress on those elephant bones and makes the elephant, as large as it is, more fragile than a cat. That might be why, no matter how fast it moves, an elephant always keeps at least one leg on the ground. The impact from landing after being in midair might be too much.

But there are peculiarities about the elephant's legs that suggest we shouldn't rule out the possibility of jumping. For instance, most four-legged animals switch through a set of different gaits as they accelerate. Typically, the animal starts by walking, then accelerates to trotting, pacing, and then galloping. By the time most of those animals are galloping, they spend a significant amount of time with all four feet in the air.

Did You Know . . . Human sprinters spend most of a 100-meter race in the air, with both of their feet just above the track.

But elephants don't seem to switch gears like that. Instead, they just keep walking faster and faster. Elephants can get up to about 15 miles per hour, but even at that top speed, they're keeping at least one foot on the ground. At high speeds, the elephant's front legs are essentially still walking: it's called the "vault" pattern, because the leg is planted and the animal moves over it in the same way that a pole-vaulter plants his or her pole and flies over it. But its hind legs are bouncing a little on impact and behaving more elastically than you'd expect from an animal of this size.

Other massive animals, such as rhinos, gallop like regular four-legged animals, but even rhinos are actually quite a bit smaller than elephants, so it might be that elephants are just too big to gallop.

There's another difference between the fore- and hind limbs: typically, quadrupeds use their back legs to push themselves forward, while their front legs are the brakes. But Dr. John Hutchinson of the Royal Veterinary College in the UK has been studying elephant locomotion for decades, and he's shown that elephants can use any leg for both braking and propulsion. Elephants are nature's 4x4s.

It's not just the force of body weight on the bones that affects an elephant's ability to jump. The other parts of the leg play an important role, too. Good jumpers have flexible ankles, strong calves, and sturdy Achilles tendons, and there's little to show that elephants have jump-worthy versions of any one of those. Most animals that can jump well do so to evade predators, but elephants don't really have any. Elephants' musculature has evolved to propel them forward, not upward. But even if an elephant isn't built for jumping, could it leap if it was forced to?

No one's ever seen it, and they might not have the muscular power in their lower legs to propel themselves off the ground. And even if they did launch themselves, the impact forces of takeoff and landing would be huge, all of which gives good reason to doubt that elephants could—or even would—jump.

History Mystery

Did Galileo drop balls from the Tower of Pisa?

A popular legend relates how, sometime in the late 1580s, Galileo climbed the Leaning Tower of Pisa carrying a musket ball and a cannonball. He was there to wage war—not on any army, mind you, but on two thousand years of scientific belief.

Up to that point, much of physics had been based on the teachings of Aristotle. One of the ancient Greek's beliefs was that objects fell in proportion to their weight. Aristotle claimed that if a 100-pound lump of clay and a 1-pound lump of clay were dropped at the same time, the bigger lump would fall 100 feet for every 1 foot the smaller one fell. Seems ludicrous now, but that was the prevailing belief then.

Galileo didn't set out that day to prove that anything of any size, weight, or density would fall at the same speed—he didn't believe that. What he did believe was that if he dropped a musket ball and a cannonball together, they would both hit the ground at virtually the same time. As a group of students, teachers, and other scientists (then called philosophers) watched from the ground with bated breath, Galileo supposedly dropped the two balls over the ramparts of the tower. It was a long way down to the ground, but the story goes that both balls hit the grass simultaneously, a brilliant confirmation of Galileo's claims and an utter rejection of Aristotle's theory, which had held sway for close to two thousand years.

If it happened today, tweets, Instagrams, and animated GIFs would dominate social media. There'd be video footage. And there would be no doubt about how the events took place. But this was more than four hundred years ago. There was no

photography or even an artist to sketch the event. Actually, nobody at the time even wrote about what happened. It wasn't until sixty years later—twelve years after Galileo died—that the experiment was recounted for the first time. The entire legend was captured in a few scant lines written by Galileo's close friend and biographer Vincenzo Viviani. Viviani suggested that Galileo had performed the experiment more than once, and several writers since have embellished the story to the point that they claimed the onlookers watched with horror as their cherished faith in Aristotle was demolished before their eyes. (The YouTube videos would certainly have included interviews with them.)

Considering the intensity of the arguments waged over whether the experiment at Pisa even happened, it's fair to ask: Who cares? After all, it doesn't matter how exactly Galileo arrived at his findings, just that he did so. But there's good evidence that at least two others before Galileo might have beaten him to the punch. In the seventh century, a little-known individual named John Philoponus said that if you dropped two objects of very different weights, they would fall, not in the ratio of those weights (the Aristotle line of thinking), but at the same time. According to Philoponus, even if one of the objects was double the weight of the other, the difference in the rate of their fall would be imperceptible.

Later, a mere four years before Galileo was said to have climbed the Tower of Pisa, a mathematician and engineer named Simon Stevin, in Delft, the Netherlands, described how he had dropped two iron balls—one ten times heavier than the other—onto a plank thirty feet (nine meters) below. Stevin reported that the two balls hit the plank so simultaneously that it sounded like a single impact.

Galileo himself never breathed a word about demonstrations at the Tower of Pisa (except, allegedly, to Viviani). He did, however, claim to have "made the test" with a cannonball and a musket ball, after which he began to predict what would happen if he dropped other objects. For instance, he predicted that if a lead ball and an ebony ball were dropped from a height of about 330 feet (100 meters), they would be no more than about a finger length apart when they hit the ground. (Note: he didn't say that he had actually tried it.)

Despite his celebrated (purported) demonstration and the fact that Galileo dragged science out of ancient Greece and into the seventeenth century, one of the ironies of his work is that his predictions about the falling objects were incorrect. Galileo neglected to consider air resistance. Objects falling through air experience drag caused by friction. The greater the velocity, the greater the drag. If something is falling from high enough, such as a parachutist, that person will accelerate until air resistance increases to equal the force of gravity, and at that point the parachutist will have reached terminal velocity.

The degree to which air resistance affects a falling object depends on both the size and density of that object. From the 1960s to the 2000s various attempts were made to re-create the Pisa experiment.

In one, an iron ball and a rubber ball of the same size (but obviously very different densities) were dropped 125 feet (38 meters)—the equivalent of a thirteen-story building—and when the iron ball hit the ground, the rubber ball was about 23 feet (7 meters) behind it. Scientists have also calculated what really would have happened with Galileo's conjectured lead versus ebony: if both balls were about 4 inches (10 centimeters) in diameter and were dropped 330 feet (100 meters), upon impact they would be 16 feet (5 meters) apart, not a finger length, as Galileo had predicted. Ebony and lead are different densities, but even if you take that out of the equation, the great scientist was still off. Galileo had theorized that if two iron balls one weighing a pound and the other 100 pounds, were dropped a little more than 200 feet (60 meters), there would be only 2 inches (5 centimeters) between they when they hit the ground. The reality is that they would be more than 4 feet (1.2 meters) apart.

In the absence of air friction, all kinds of combinations of sizes, densities, and shapes of objects would fall at identical speeds and land at the same time. That's been proven by performing tests in a vacuum. While it's possible to create a vacuum on Earth (and the Galileo experiment has been duplicated in one), it's far more spectacular to test the theory on the moon. When the Apollo 15 astronauts were there, astronaut David Scott performed the experiment with a feather and a hammer. He dropped both at the same time, and the grainy video of the test shows both objects landing on the moon's surface at exactly the same time. Although the astronauts were confident of the results they would get, there were apparently risks that the experiment wouldn't work. Before the successful trial, Scott had tried it once, and static electricity caused the feather to stick to his hand. Scott's bad luck revealed another important issue involved when trying to drop two objects of very different weights simultaneously while on the moon or anywhere: you tend to let go of the lighter one first, especially if you've been holding them at arm's length.

TRY THIS: Cut out a paper disk the same size as a quarter. Put the paper disk on the tip of one forefinger and an actual quarter on the tip of the other. Now drop both of them simultaneously. The coin will reach the floor before the paper disk. From this experiment, it is possible to conclude—mistakenly—that heavier objects fall faster.

Now place the paper disk on top of the coin and drop them together. Both will reach the floor at the same time. Why? By putting the paper on top of the coin, you've shielded it from the air resistance it encountered in the first experiment. The second experiment shows that it's not the mass of the object that causes it to fall faster or slower but the resistance or air friction.

It's possible that none of these facts have convinced you and that you still think that heavier objects fall faster than lighter ones. It's an intuitive conclusion, but intuitive physics leads people astray all the time. After all, about a quarter of North Americans believe that if a person runs off a cliff, they'll move straight out until they slow down and stop, at which point they'll fall straight down à la Wile E. Coyote. If you're of that mind-set, then consider this thought experiment. Pretend we have a ten-pound ball and a one-pound ball. Let us assume, à la Aristotle, that the ten-pound ball falls faster than the one-pound ball because it's heavier. Now, let's attach the two balls. What happens next? A traditional Aristotelian might think that the combined object should fall slower, since the slower movement of the one-pound ball would hold back the ten-pound ball's descent. But by the same group's reasoning, the two balls should fall faster when they're attached because they weigh more, and heavier objects move faster. It's impossible for both outcomes to happen, so the only possibility is that they were falling at the same rate in the first place. Chalk up another point for Galileo.

It's very unlikely that further documentation will surface to shed definitive light on the incident at Pisa. So why, as I asked before, is something so apocryphal so celebrated? In a single, dramatic story, one man completely reversed millennia-old thinking about how things work. It's not the science; it's the showmanship!

That's why.

Part 4
The Natural World

What's inside a black hole?

To understand black holes, we first need to understand the life cycle of a sun. The beautiful thing about our sun is its consistency: just as it sets tonight, so it will rise tomorrow, and each day it will shine with the same intensity. Thanks to the crushing force of gravity keeping it in check, the sun—our closest star—has been a giant nuclear fusion reactor for billions of years and will continue to shine for about five billion more years.

But all good things come to an end. Eventually the last of the sun's hydrogen will get used up, and its upper layers will push outward. The whole thing will heat up, swell, and then shrink again. A dying sun goes through cycles like this until it sheds all of its outer layers, leaving behind what's called a white dwarf, the dense leftover of the sun's core, or half the sun's current mass packed into an object the size of the earth.

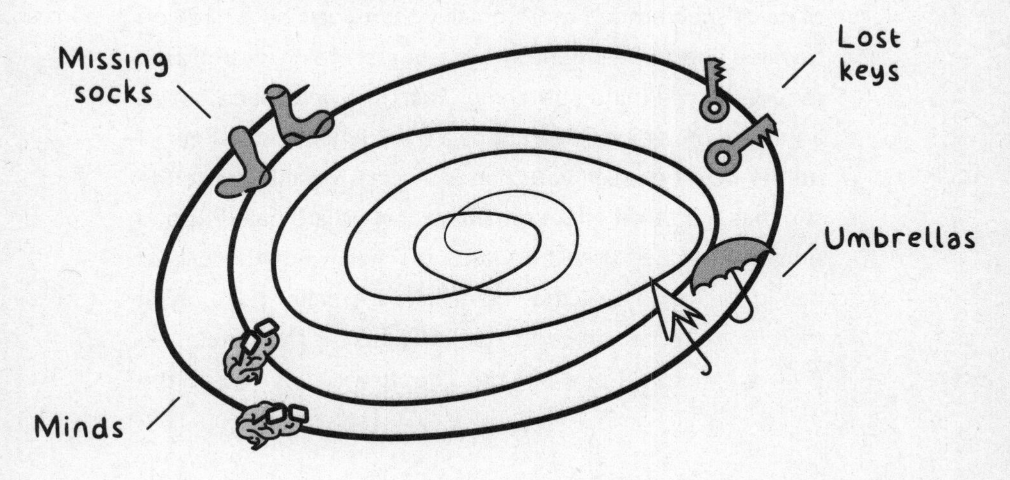

The bigger the star, the more dramatic this dying process is. But if this story begins with a star that is even bigger—say, twenty times or greater than the mass of our sun—it will still shrink as it runs out of fuel, and the shrinkage never stops. The gravitational force of that much mass will simply continue to collapse until it reaches an incomprehensible state of infinite density packed into an infinitely small space. This is a black hole.

There are black holes that are star sized, some that are supermassive and, theoretically, mini black holes. Each one has the same mass it did before it collapsed (minus any material that was shed into space), so there must be something in them. Black holes, especially the huge ones, vacuum up stars and gas that they encounter. As that material swirls down the drain of the black hole, it heats up and emits huge quantities of X-rays and radio waves. From the pattern of those objects, astronomers can determine the size of the black hole that they're orbiting. The late physicist John Archibald Wheeler described this process as similar to standing in a poorly lit ballroom where the male dancers are dressed completely in black but their female partners are in white. You would be able to make out only the female dancers, but you could judge by the movements of the females where the male dancers were.

TRY THIS: A black hole's closest relative is a wormhole. Wormholes are both exits and entrances in space—places that you can (hypothetically) move through to reach distant parts of space much more quickly than you could had you traveled the conventional way. Understanding wormholes requires some mind stretching. First, imagine space not as a three-dimensional area but as a two-dimensional sheet of rubber held between your hands. A massive object, like the sun, has the same effect on space as a billiard ball dropped onto the rubber sheet: it warps it. Then, with the billiard ball in the middle of the sheet, take a smaller ball—say, a marble—to represent earth, and roll it across the sheet.

It won't travel in a straight line. It will gravitate to the middle and travel around and around the billiard ball (representing the sun) in an orbit because of the slope of the sheet. Now imagine using a pen to make two ink marks at opposite ends of this rubber sheet. Now fold the sheet to position the two points on top of each other. Once the points overlap, you can either travel from one to the other by sticking to the surface of the sheet and tracing a path all the way around, or, instead, you can punch your way right through the rubber: that's the wormhole. Congratulations, you just traveled in hyperspace. Just don't make the mistake of thinking you can travel through a wormhole by entering a black hole: that would be the end of you!

We can't know anything about a black hole beyond what's called the event horizon. This is the border surrounding the black hole. Once you cross it you can't ever come back, because the black hole's gravitational force is so strong. If you did cross it, your experience would be very different from those watching you. Outsiders tracking you as your spaceship descended toward the event horizon would see you get closer and closer, and then you would appear to stop. The light of your image wouldn't be sucked into the black hole, but it wouldn't be able to escape the force of the hole's gravity, either; it would remain there, suspended in space against the darkness of the black hole.

As the person flying into the black hole, though, you would experience events much more quickly. What makes a black hole special is not

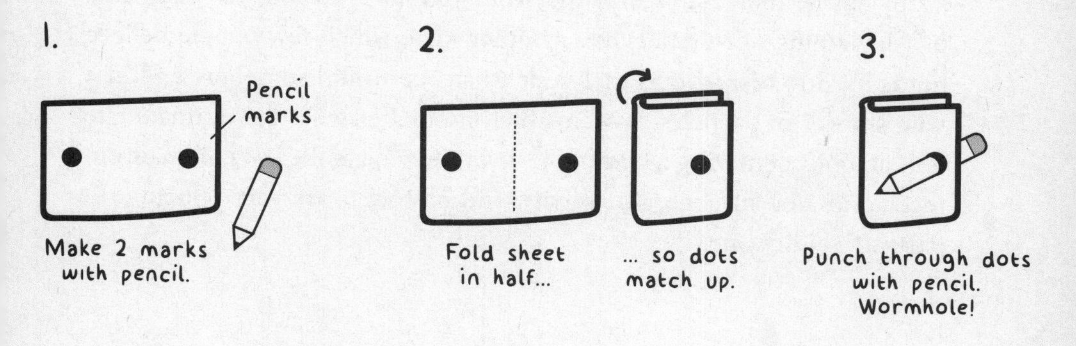

that its mass is unbelievably huge but that the mass is packed into such a tiny space—a single point—and so its gravitational effect is much more intense. On the earth, because your feet are slightly closer to the center of the earth than your head, you are technically experiencing an infinitesimally stronger gravitational force on your feet. But the distance between your head and feet is so small relative to the distance to the center of the earth that you don't feel the effects.

A black hole is a much more concentrated source of gravity, not to mention more massive than the earth. As you get closer to the black hole's center, you would begin to be stretched. That can't last for long, of course, because you'll start to break apart, first into a top half and bottom half, then into four, then eight, and so on. At the same time, space-time itself is being channeled into the black hole, meaning that you're being squeezed from the sides, as Neil deGrasse Tyson likes to put it, "like toothpaste." It's actually more like the last gasp of the toothpaste when it comes out in splatters—that's what it would be like for you to enter the black hole. This simultaneous stretching and squeezing has a technical term: spaghettification.

After you enter the black hole in skinny little pieces, it's unclear what exactly would happen, because we have no way of looking inside. The "black" in the name means that no light escapes from the intensity of the hole's gravity, and where there's no light, we can't see what's going on. What we know for sure is that the laws of physics (as we know them) no longer hold. Maybe space simply ends there and the story's over. But as far as we know, matter shouldn't be able to end its existence, so that doesn't seem right. Although there's an infinite density squeezed into a tiny space, there's a chance that your remnants—even the fragments of your atoms—could survive. Another idea, which few people believe but is hard to disprove, is that your existence would somehow end in a wild shower of particles. It seems that, no matter what the ultimate fate is of an object entering a black hole, it would violate the laws of quantum mechanics and/or general relativity, and physicists are very reluctant to part with either one.

Why is the night sky dark?

That seems like an easy question. Even on the most spectacular nights for stargazing, most of the sky is dark. A star here, a star there, but pinpricks—not enough to light up the night. It's so dark that intergalactic space is more than a million times dimmer than the light you're reading by.

At first glance, it makes sense that it would be dark in space. But something's missing. Stars give off light. And there are billions of them. The number of stars in our Milky Way Galaxy alone is probably 100 billion. And that's just our galaxy. There are hundreds of billions more galaxies out there, as the Hubble Space Telescope has shown us. That guarantees that wherever you look in the night sky, you can be sure that you're staring at a wall of stars, or, if you prefer, a wall of galaxies.

The physics says it doesn't matter whether all the stars are clumped into galaxies or are off by themselves; it should still be bright out there. The

physics also concludes that any patch of the night sky, no matter where you look, should be about as bright as the surface of the sun. So there's a huge puzzle here. It's called Olbers' paradox. There are two theories that carry some weight when astronomers argue about Olbers' paradox.

One theory attributes the darkness to the expansion of the universe, which has been going on since the big bang. And it isn't like everything is flying apart as it does when there's been an explosion in the kitchen. Instead, space itself is expanding.

Because of that expansion, the space between the galaxies is increasing, with the result that galaxies are receding from us. As that happens, the light from the galaxies undergoes a physical shift that we are familiar with from sound: the Doppler effect. Listen to the whistle, siren, or sounds of the engines as a train or ambulance nears. As it approaches, the sound is high-pitched; as it recedes, the sound is low-pitched. It happens because sound waves are either pushed together (higher pitch) as the vehicle approaches, or stretched out (lower pitch) after it's passed.

The same happens to light. When a light source is receding at rapid speed, the entire spectrum of visible light is stretched à la the ambulance in the previous example. And when light waves are stretched enough, they exit the visible spectrum to become infrared light, microwaves, or even radio waves, all of which are invisible to us. Even so, this effect dims the light only by about a factor of two, not enough to make the night sky as dark as it is. Also, some scientists have pointed out that shifting of light due to high speeds should also happen at the other end of the spectrum, where invisible waves of ultraviolet light should stretch and become more visible rather than lost. So the mystery remains.

The other major factor that relates to darkness in space is the age of the universe: 13.82 billion years. No galaxies are older. In fact, the first stars began to shine about 200 million years later than that date and the earliest galaxies somewhat after that. That means that no galaxy that we can see can be any farther away than 13.82 billion light-years: that's the maximum distance that light from that galaxy could have traveled in that time. Because the speed of light is finite, there will be galaxies farther away whose light hasn't had time since their births to reach us. It's also true that galaxies die and wink out, and that new galaxies are coming to life all the time, but the continuing and even accelerating expansion of the universe guarantees that many will always be out of range. Imagine there's an opaque dome roughly 13 billion light-years away—we will never see anything on the other side of it.

The paradox of "so many stars, so little light" has attracted big thinkers over the years. Famous comet man Edmund Halley wrote about it; eighty years later, Wilhelm Olbers donated his name to it and came up with a theory that light was being intercepted by material between the stars. Unfortunately, he failed to give credit to the man who anticipated his ideas, but they were wrong anyway, because any material that did intercept starlight would heat up and eventually become hot enough to radiate light itself.

The most unexpected contribution, though, came from Edgar Allan Poe, the poet, short story writer and, apparently, cosmologist. In "Eureka: A Prose Poem," published in 1848, he anticipated the importance of the age of galaxies. He argued that the only way to explain the darkness was to imagine that the invisible background was so immense that no ray from it had been able to reach us.

Poe's insight was pretty amazing, considering how little was actually known about the universe at that time. As it turns out, he was right: the sky, as seen from earth, will always be dark because most of the light that's out there is too far away to reach us.

Why is the sky blue?

Even with this straightforward question there are subtleties, but the main part of the answer is pretty clear: sunlight, which is white, doesn't travel through the earth's atmosphere unscathed. When sunlight collides with molecules of air, it's scattered in all directions. But shorter wavelengths of sunlight are more susceptible to scattering than longer ones—so light at the violet and blue end of the spectrum scatters more than red, yellow, or green.

If the sun is in the east and you look at the sky in the west, the blue you see is the light that has been scattered away from the sun and then toward your eyes. However, if you look in the direction of the sun at sunrise or sunset, the sky looks red or orange. That's because the sun is near the horizon and you're looking through much more atmosphere and see only the small fraction of the sun's spectrum that's most resistant to scattering: red and orange.

So why don't we see the sky as violet? After all, violet has shorter wavelengths than blue and should be scattered more intensely by the atmosphere. There are two factors that rule that out. One is that the violet part of the sun's spectrum is less intense than the blue, so there's simply less violet light. The other is that our eyes aren't as sensitive to violet as they are to blue. But, if that's so, how can we see violet in a rainbow? Because the violet band there is separated from the other colors, whereas in a sunny sky it isn't.

There is a truly peculiar aspect to the blueness of the sky, and it's this: it's not even clear that we were always able to see the color blue. Or at least, if we could see it, we were so unaware of it that we didn't give it a name. This is, as you might guess, a controversial topic!

It starts with the ancient Greek poet Homer, the author of *The Iliad* and *The Odyssey*, two epic poems thought to have been written about 2,700 years ago. There have been many studies that have focused on these two books, but surely one of the weirdest is the one counting the number of times different colors are mentioned. The score is this: black 200, white 100, red fewer than 15, yellow and green fewer than 10, and blue: 0. Zero! The sea, according to Homer, was "wine-dark," not blue. Oxen the same. In one instance, sheep were "violet."

This Homeric study sparked interest in the subject, and subsequent surveys showed that many ancient languages also lacked words for blue and lacked names for other colors as well. In fact, there emerged an apparent cross-cultural sequence in the way color words first appeared in languages: black and white first, then red, then yellow and green, and finally blue.

So what's the explanation? Homer was a poet (except "he" was more than likely many different people); color words could simply have been the artist's choice. But the evidence that other languages only gradually incorporated color words makes it more interesting. There has also been a handful of experiments that demonstrate that different cultures appear to see, or at least to label, colors differently. In one, an African tribe, the Himba, could tell minuscule differences among shades of green that to most of us would look absolutely identical, and yet couldn't pick out the one square of blue sitting among the green.

Why the hierarchy of black and white, then red, then yellow, and last—and least—blue, though? One suggestion has been that we don't need labels for colors unless we work with them. So the ancient Egyptians did have a word for blue, but they were also one of the very few ancient cultures, if not the only one, to use a blue dye. Use it, name it! This might account for the relative popularity of the word for red, a common dye and, of course, the color of blood.

Another reason offered for the rarity of words for blue is that there are few things in nature that are blue. No blue plants, very few naturally occurring blue flowers, no blue animals. Most Mediterranean or Middle Eastern peoples of ancient times weren't blue-eyed. But in North America we have birds: blue jay, blue grosbeak, blue-winged teal, bluebird. And, of course, the sky.

No one is suggesting that somehow our vision has actually changed over the last couple of thousand years. Instead, the thought is that words aren't necessarily attached to colors unless those colors acquire some importance. So how would Homer, or any other sages of the ancient world, have described the sky above them? Sure, it's sometimes leaden, sometimes even gray or black, or red and orange at sunset and dawn. But at midday?

I don't know, but I can't leave the topic without pointing out that in Samuel Butler's 1900 translation of *The Odyssey*, there's the "child of morning, rosy-fingered Dawn" (that fits), "the gray sea" (black combined with white) but also "the deep blue waves of Amphitrite." Aha! I likely haven't really discovered anything of importance, given that this is a twentieth-century translation of the ancient Greek. Butler simply may have interpreted Homer's words as signifying blue, if not doing so literally.

Why does the moon look larger at the horizon?

If you look at the moon when it is just rising or setting, it looks almost twice as big as it does when it's overhead. Although people have seen this happen with every full moon for millennia, it has been extremely difficult to come up with an explanation for this illusion.

Early stargazers suggested that the moon actually changed size. They believed it inflated during the day, like a balloon, out of our sight, then rapidly shrank as it rose in the night sky, only to re-inflate just before dawn. But that theory is (justifiably) long gone.

It is true that because the moon's orbit is more of an oval than a circle, at any given point in its orbital cycle its distance to the earth can vary by more than 24,000 miles (40,000 kilometers). The change in the distance between the moon and earth, however, takes more than a single night to happen. So that doesn't explain why the moon looks so much bigger at dawn or dusk.

On to the next: people reasoned the moon might appear larger at the horizon because it was closer to the earth then. The opposite is actually true, although by an unimportant amount: when you look at the overhead moon, you're standing on the closest point on the earth to it, but the horizon moon is literally half an earth (roughly 3,700 miles, or 6,000 kilometers) farther away.

Another possibility was then considered: that the atmosphere somehow bent the light that reflected off the moon. But you can quickly disprove that idea with photographs. Sequential pictures of the moon taken over the course of a night show that it's exactly the same size at all times—that is, the size of the moon in a photograph will always be the same no matter where in the sky the moon is when you take the picture, unlike the apparent dramatic changes we see when we're looking at it with our eyes.

TRY THIS: You can do a simple test to check the size of the moon and see for yourself whether the change in size is an illusion. Take an aspirin outdoors when there's a full moon, and hold it up at arm's length to cover the face of the moon. First, you'll be astonished to see that a tiny aspirin can cover that giant horizon moon, and you'll be even more astonished to see that later that night, when the moon is overhead, it will still just cover it. Next, observe the horizon moon through a toilet paper roll. The moon will suddenly appear smaller than it did when you were looking at it without the cardboard tube. What the photograph and toilet paper roll experiments show is that the differing sizes of the moon appear to have something to do with how we view it. More accurately, the difference has to do with how we think we see the moon. It's all an illusion.

Although the moon's varying sizes are an illusion, distance still plays a key role in how it appears to us. Your distance to the moon is more or less the same no matter where the moon is in the sky. You do get closer to the moon as the earth turns, but as explained earlier, that distance is trivial (relatively speaking). What's more important than the distance itself is how far that distance seems to our brains. One theory is that the horizon moon seems much farther away than the overhead moon because you can compare the horizon moon to everything on the ground between you and it—such as buildings, mountains, or trees.

The problem with this theory is that we don't really need any intervening objects to experience the moon illusion. Pilots at high altitudes have experienced the bigger horizon moon, even with no other large or moving objects to compare their view against. The moon illusion also works over deserts or water—places where there are few, if any, geographical features.

TRY THIS: On a clear evening, turn your back on the horizon moon, then bend down and look at it from between your legs. You will see that the moon appears to be smaller from this awkward vantage point. You can even try an experiment within this experiment: How far upside down do you have to be before the moon looks small?

So what is it that makes the moon look big near the horizon? We know that the image the moon leaves on your retina is the same all night, from moonrise to moonset; its size never changes. One of the most popular explanations is that your brain doesn't imagine the sky to be the inside of a sphere, but rather, more like a shallow inverted bowl—distant at the horizon but much closer overhead. So your brain has two conflicting inputs.

On one hand, the image of the moon on your retina is the same at every point, but your brain has to reconcile that with its assumption that the horizon moon is actually farther away. The only way two objects at different distances away can leave equal-sized images in your eye is if one is bigger. And so it appears that way.

What would happen if the moon disappeared?

The moon's pockmarked surface is evidence that over its lifetime it's been smacked by untold numbers of space rocks. Chunks of rock are hitting the moon all the time and we don't even notice, although we might have once, back in 1187. In June of that year monks in Canterbury, England, reported seeing the crescent moon split in half by fire. This might have been a piece of space rock, but it should have left a huge crater behind, and none has been found that convinces astronomers that it's the one.

Although that collision apparently didn't spark a meteor shower, more than a hundred meteorites found on the earth have come from the moon. There's no complete count of craters left by such impacts, but it would likely number in the millions.

Science Fact! *At the time of its formation, the moon was much closer to earth than it is now: its orbit took just 6 days, not roughly 28. The full moon would have been an incredible sight—16 times bigger than it appears now—looking the size of a saucer, rather than the diameter of a pill held at arm's length! And every few days there would have been a total eclipse of the sun.*

The moon is massive, and a similarly massive asteroid would be needed to break it apart or even nudge it out of its current orbit. As far as we know, there aren't any asteroids in our solar system like that, let alone on an orbit that would result in a collision course. So the moon will stay more or less where it is. I say "more or less" because it's slowly retreating from us—not enough that you'd notice, but each year by 1.49 inches (3.78 centimeters), about the same amount as our fingernails grow in the same time.

Why is that? Ever since the moon was created by massive collisions of space rock 4.5 billion years ago, it has been slipping away from us, because of the tides—not just the ones we're familiar with in our oceans but the reverse: tidal forces exerted on the moon by the earth. Because the earth rotates faster, it drags its tidal bulge slightly ahead of the moon, and the gravity of the mass of that bulge tugs the moon with it. That slows down the earth's rotation and at the same time speeds up the moon slightly. As that happens, the moon shifts to a higher orbit.

What if the moon actually did disappear? What difference would that make? If you think of the earth as a spinning top, right now it's tilted about 23 degrees from vertical. It wobbles over tens of thousands of years, but not wildly. The moon's powerful gravitational influence is crucial in stabilizing that tilt. If the moon suddenly vanished, it's estimated the tilt could range from 0 degrees (straight up and down) to 85 degrees (almost lying on its side). Since the tilt of the earth is responsible for the seasons, and the slow wobble is a major factor in the periodic ice ages, a major change would provoke catastrophe.

Did You Know . . . There's speculation that life on earth might have originated in tidal zone areas, in that era long ago when the moon was closer and the tides were more far-ranging than they are now.

Those dramatic tides in the Bay of Fundy and the amazing surfing in Hawaii? Pretty much over. Sure, the sun exerts some tidal forces on the earth, so tides wouldn't disappear entirely, but they'd be nothing like they are now. That would have a dramatic effect on thousands of life-forms, some of whose life cycles are precisely tuned to the lunar calendar. And I've already mentioned those millions of craters on the moon created by impacts by space rock. With no moon to intercept them, we would be exposed to the same bombardment.

And what about us? If we aren't killed by dramatic climate change or asteroids, our lives would still be altered. Stargazers irritated by moonlight washing out the rest of the universe might be pleased (although they'd miss their eclipses), but no more moonlit nights for the rest of us. The lack of a moon would also disappoint romantics and writers of cheesy songs. No more conspiracy theories about whether astronauts really went to the moon or not. No more howling at the moon, nothing for the cow to jump over, and certainly no more strange human behavior during the full moon.

But wait! There is no strange behavior during the full moon, despite what you've heard. No more crime, no more hospital admissions, no higher rate of births, no more suicides or homicides—nothing. Rumors persist, but studies reveal the truth: the full moon causes no apparent change in human behavior.

How do stones skip?

When it comes to skipping stones, there are three kinds of people: those who love it as a cottage activity, those who take it seriously enough to try for world records, and those who try to understand the physics of it. It's incredible that there are still physicists focused on this question, because people have been skipping stones for ages.

The ancient Greeks referred to skipping stones across water millennia ago. But the first well-attested records are from the early seventeenth century, when King James I of England, who ruled from 1603 to 1625, amused himself by skipping gold sovereigns across the river Thames (he was king, after all). Gold sovereigns weigh about 8 grams, or a little more than a quarter of an ounce. That's extremely light, and as anyone who's skipped a stone before knows, when a rock is too light, it doesn't stay horizontal long enough to skip well. I bet if King James was somewhere in central London, he was throwing at low tide, making it easier to get the coins across the river.

The first scientific account of stone skipping comes from the lab of Lazzaro Spallanzani in the 1700s. Spallanzani didn't just watch people skip stones and guess what was happening, though. Instead, he dropped and threw stones and recorded the results. He even managed to get stones airborne by grasping them between thumb and forefinger and sweeping them along the water. Some of his observations were pretty straightforward: for instance, he showed that you need a flat stone and that it has to hit the water on its flat side, not its edge, if it's to skip. He then added the crucial observation that a stone will skip much higher if it hits the water at a slight upward angle.

Did You Know . . . Spallanzani was an accomplished guy. Not only was he the first to show that bats echolocate, he also did crucial experiments to disprove the concept of spontaneous generation—that life could simply erupt from non-living matter. He showed that broth left exposed to the air quickly spoiled and became infested with bacteria, but that the same broth sealed off from the air did not. No surprise to us, but a revelation at the time.

Most of those points are obvious to anyone who's skipped a rock before. But Spallanzani dove into the details. He fired a lead shot at a low angle across the water. From what he witnessed, he argued that when a stone thrown (or a bullet fired) at a low angle strikes the water, it creates a depression, rides down the near side of the divot and then slides up the other side and into the air. It's an astonishing observation, because it would have required a ton of patience and a keen eye.

Spallanzani's findings weren't referred to again until the twentieth century, when modern technology allowed for more precise measurements. In a high-speed video shot in the late 1960s, researchers were finally able to view in detail what Spallanzani had observed with his naked eye. The video showed that a skipped stone does indeed hit the surface and push down the water, creating a wave in front of it. The rock then continues to skim forward along that wave, the front edge slanting higher and higher until it reaches an angle of something like 75 degrees (for comparison, the Tower of Pisa stands at an 80-degree angle). At that point it finally tears free of the water and launches itself into the air again, where of course gravity immediately begins to pull it back down.

The angle of throw is crucial. Lydéric Bocquet, a physicist at the École normale supérieure in Paris, has established that, ideally, the stone should hit the water at a 20-degree angle—that is, the leading edge of the rock should be 20 degrees above horizontal, about the same angle as a modest waterskiing ramp. That impact angle of 20 degrees minimizes the amount of time that the stone spends in contact with the water, which is crucial, because the water exerts a powerful drag on the stone, slowing it down. In general, the water is a thousand times denser than the air, so the more air time the better. A perfectly skipped stone will spend about one hundred times longer in the air than in the water.

The stone has to be thrown fast, but more important, it has to be spun fast: spinning keeps the rock stable when it hits the water. Thanks to something called the gyroscopic effect, the spinning helps prevent the rock from wobbling to one side or the other when it strikes the water's surface. Bocquet figures that a stone spinning five times a second will skip five times, but it has to spin almost twice as fast to reach fifteen skips.

TRY THIS: The physics behind skipping stones is one thing, but what technique is best for getting that perfect 20-degree angle and spin? Start by facing the water sideways and bending your knees. Hold the rock between your thumb and your first finger. Bring your arm back and keep the flat side of the rock parallel to the water as you throw it. At the last second, snap your wrist to flick the rock against the surface of the water.

Of course, even if your technique is perfect, every time the stone hits the water, it will give up some of its energy due to friction. Eventually, all good things must end. The final stages of a stone skip are marked by shorter and shorter skips, followed by a brief period when the rock waffles through the water without leaving the surface. These not-quite-skips are sometimes called "pitty-pats" by experts, and soon after the rock reaches that stage, it sinks.

There are annual rock-skipping competitions all over the world, and the numbers from them are startling. The most consecutive skips, as recognized by Guinness World Records, is an astounding eighty-eight, set by Kurt Steiner at Red Bridge, Pennsylvania, in September 2013. Steiner held several records before that particular toss and apparently spends much of his time looking for the perfect stones, which he describes as those that weigh about five ounces (King James I, take note!), measure close to one-quarter-inch thick, and are extremely smooth on the bottom.

But forget the quality of the stone—the mechanics of a throw like that are extraordinary. According to Bocquet's equations, that stone had to have left Steiner's hand traveling at something like 18 meters per second, or about 40 miles per hour. Of course, this assumes a number of things about the precise angle of the throw, the actual weight of the stone, and so on. But still, it would take great strength to get the rock moving that quickly, so it seems that Steiner lived up to his nickname: "Man Mountain."

As always, scientists do what scientists do. They've taken a casual activity and standardized the process by skipping perfectly round or square discs of sandstone or acrylic. That reinforces what stone skippers have figured out intuitively, but I'm with Kurt Steiner: when it comes to the proper way to skip a rock, patrol the shore, look for the perfect natural stone, and let it fly.

Why does campfire smoke seem to follow you around?

Many people have argued that when you move around a campfire, you leave a partial vacuum behind that sucks smoke toward you. But nature truly abhors a vacuum, and creating even a momentary partial vacuum is extraordinarily difficult. So how does smoke seem to always know where you are?

The real answer has to do with airflow. Typically, cool air from beyond the perimeter of a fire flows toward the flames, warms, and then rises. As long as there is no disturbance in the air, there is an even and steady airflow. But if you change position, or even worse, stand up and start moving around, you disturb that air cycle, creating an eddy in front of you. This always happens when a fluid (in this case, the air) meets an obstacle (your body). It's not unlike the effect of a rock in the middle of a stream or dead leaves scattered in the wake of a bus.

Let's say you're facing the fire. The incoming air reaches your back

and is channeled to either side of you as it flows past. But the two streams don't stick to your sides and immediately rejoin at your belly button. They flow straight past, leaving an empty space in front of you that needs to be filled. As the two streams of air near the fire, they reverse direction, flowing back toward you. As they move back toward you, they carry the smoke from the fire with them. The next thing you know, you're enveloped in a cloud of campfire smoke. Not only that, but the bigger you are, the more smoke gets drawn toward you—the bigger the obstacle, the more air gets displaced and pulled back into the eddy. Change places and you'll simply trigger the same effect wherever you go.

Another thing that might affect how smoke follows you is the design of the fire itself. Adrian Bejan of Duke University has calculated that the ideal shape for the logs in a fire is a cone or pyramid, where the height is equal to the length of the base. Bejan argues that a cone perfectly balances airflow and geometry to generate the most heat for those sitting around the fire; basically, the fire gets hotter because it can breathe better.

But there's another benefit to this geometrically ideal campfire. Its height and shape allow a smooth smoke plume to form and rise more or less straight up from the core of the fire. If the smoke rises from the center, it is less affected or disturbed by motion on the edges of the fire pit, and so it's less likely to surround you as you move.

Of course, that's all presuming that you're standing around the fire in ideal conditions. Once you throw in wind, weather, and other people moving around, you have a perfect storm of physics in which all sorts of things can happen. Some campers advise placing a large stone or stump—one that surpasses you as a source of turbulence—on one side of the fire to act as a permanent obstacle. Bejan's reaction to that idea is that it had better be a big enough stone—in his words, "an obelisk." Or you could just make sure you're not the biggest person standing at the fire.

How are champagne bubbles different from beer bubbles?

All bubbles have a life that is—by human standards, anyway—extremely short. Their deaths are even shorter. Bubbles in both beer and champagne contain pressurized carbon dioxide (CO_2). During the fermentation process of both drinks, yeast breaks down sugar, generating, among other things, alcohol and carbon dioxide. As the carbon dioxide is created, it dissolves into the beer or champagne, and when the bottle is sealed with a cap or cork, the gas is trapped inside. Some carbon dioxide collects in the air space just under the cap or cork, but the rest remains dissolved. They're in balance as long as the cap or cork is in place, and so bubbles will not form. But when the seal is broken, the pressure in the air space

drops and the gases, including the carbon dioxide, rush out of the bottle. Now the balance is disturbed, and the rest of the CO_2 inside the drink must escape. The most efficient way for the CO_2 in the liquid to do that is in the form of a bubble.

For bubbles to form and escape, chemistry demands that there be "nucleation sites"—places where the bubbles can settle and grow—on the inside of the glass. It used to be thought that these were mostly defects in the glass surface, but further research has shown that they are more likely to be tiny stray fibers of cellulose that have settled in the glass from the air or a cloth.

Even when a nucleation site is available, though, forming a bubble is a tricky business. After a bottle is opened, CO_2 molecules rush to each nucleation site. When a bubble is first developing, it's less than 1/100,000 of an inch across, but as more carbon dioxide molecules collect together, they eventually reach a critical size, become buoyant enough and start to rise in a bubble. As each bubble floats upward, it becomes its own nucleation site, and so it collects more and more CO_2 as it rises. As it ascends rapidly to the surface of the liquid, it swells to a couple of thousand times its original size.

When a bubble reaches the surface, most of it remains submerged, but a small part of it protrudes into the air. The liquid of that exposed cap starts to drain away, to the point where almost anything—a slight tilt of the glass or bottle, even a wisp of air passing by—will burst it. When a bubble breaks, the impact causes a high-speed jet of beer or champagne to explode upward from the surface. Multiply that by hundreds of bubbles and you can't miss the popping and fizzing of the tiny alcoholic fireworks display. When you settle in for a sip, some of those flying bubble fragments bring the flavors of your favorite drink to your nose—the more broken bubbles, the more intense the aroma and flavor.

TRY THIS: Open a bottle of beer or champagne and then watch the strings of bubbles as they float to the surface. You'll notice that the bubbles in the string are farther apart the closer they get to the surface of the liquid. That's because the more CO_2 that the bubble collects, the more buoyant it becomes and the faster it moves.

There are many differences between the bubbles in champagne and those in beer. For one thing, a bottle of champagne contains three to four times as much CO_2 as a bottle of beer. Also, a bubble release site in champagne can generate up to thirty bubbles a second, roughly three times the number in beer. That explains the fun of popping the cork off a bottle of champagne in a celebration. Because there is a great concentration of bubbles in champagne—they grow faster, and the bottle is bigger than a typical beer bottle—the "eruption velocity" of champagne is one hundred times that of a bottle of beer.

Both beer and champagne are complex mixes of elements, and some of those chemicals can attach to the surface of a bubble. This happens more in beer than it does in champagne. In beer, some substances—for instance high-molecular-weight proteins and isohumulones from the hops—stabilize the bubbles, slowing their upward mobility and making them less likely to burst.

As the bubbles in beer hit the surface, instead of breaking apart, they bind together and form a frothy head. Champagne is very different: there are fewer of these stabilizing substances, and the bubbles rise much faster,

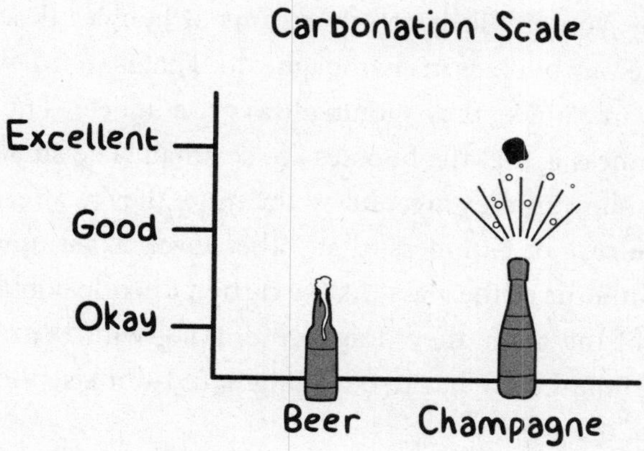

Carbonation Scale

Excellent
Good
Okay
Beer
Champagne

ensuring that champagne bubbles burst immediately upon reaching the surface.

When it comes to pouring the perfect head of foam on a beer, Guinness reigns supreme (and no, they didn't pay me to say that). The head on a pint of Guinness is smoother and far longer lasting than it is on other beers. The secret? Guinness contains nitrogen gas in addition to CO_2. Bubbles of nitrogen are different from regular old carbon dioxide: they're smaller and better at supporting the kind of dense foam that characterizes a glass of Guinness.

In the pub, a Guinness pour—or "pull"—is forced under pressure through something called a creamer plate, which is filled with tiny holes that minimize the size of the bubbles. But when Guinness began to sell their beer in cans, re-creating those same conditions was a challenge. Nitrogen doesn't dissolve willingly in beer, so most of the gas—which was introduced into the liquid during the canning stage—hung around at the top and escaped when the can was popped. No nitrogen in the liquid meant no distinctive Guinness head.

Guinness solved that problem by introducing the "widget": a tiny ball filled with nitrogen and carbon dioxide that's added to each can. As long as the can is sealed, the widget doesn't do anything. But when the can is opened, the pressure change causes the widget to release its combination of gases. The stream of nitrogen bubbles agitates the beer, creating a long-lasting creamy head.

There's one final oddity about Guinness. If you watch a glass of the stuff carefully, you are likely to see streams of bubbles descending, not ascending the way bubbles in champagne do. That is weird, because bubbles are buoyant, and so they should always rise in beer. The reason they don't in Guinness is that the bubbles are so small. The smaller the bubbles, the less buoyant they are, and so the more they're affected by other forces. In the case of Guinness, that "other force" is an upward flow of liquid in the middle of the glass. As the carbon dioxide bubbles rise from the bottom of the glass, they drag liquid along with them, and where there's rising liquid there has to be falling liquid—or else the beer would levitate out of the glass!

In a traditional pint glass, the carbon dioxide escapes up the middle of the glass, and so the nitrogen bubbles—because they're too weak to resist other forces—are pushed to the outer edges of the glass and descend along the wall. In one cool experiment, though, researchers showed that an "anti-pint" glass—one that's fatter at the bottom than the top—can reverse these currents. If Guinness is poured into an anti-pint glass, the bubbles flow downward in the middle of the glass and rise along its walls. The experimenters came to a simple conclusion: because the anti-pint glass is wider at its base, carbon dioxide bubbles that form along the bottom are crowded together by the walls as they rise. The only place the nitrogen can sink is right down the middle, between the escaping streams of CO_2 bubbles.

If you want to try these experiments yourself, Guinness is the best beer to use because the sinking bubbles are highlighted against the beer's dark color. It makes for easy viewing . . . but remember, if you tilt the glass to drink from it, you're going to disturb the bubble display.

Why do leaves change color in the fall?

Considering how beautiful fall colors can be, you might think that it takes a lot of energy for leaves to change their hue from their summer green to the reds, yellows, and oranges of autumn. But in fact the opposite is true: many of the color changes you see in the fall are the unintended byproducts of a cost-saving measure by trees.

Leaves are green because they contain chlorophyll, a molecule that collects solar energy. Chlorophyll absorbs all light across the sunlight spectrum except for green, which it reflects back to our eyes. During the growing season, when there are the most hours of daylight, chlorophyll is at peak performance, converting the energy of sunlight into sugar molecules that can release the energy later or be used to build other molecules needed by the tree. But when sunlight diminishes in early fall (it shrinks steadily from June 21 through to December 21), the cost-benefit equation for the tree changes. The energy that a leaf absorbs from the sun dwindles to less than the energy it uses to perform the chemical work required to maintain its sophisticated machinery. To prevent wasting resources, the tree begins to shut down its leaves.

As the leaves shut down, chlorophyll molecules are shipped back into the tree. (Nitrogen is a limited resource, too, so the tree also stores it for use the next spring.) As the chlorophyll is withdrawn, other compounds in the leaf are left behind. Most of those compounds are carotenoids, and two of the most important are carotene and xanthophyll. Carotene is orange (carrots), while xanthophyll is yellow (squash). As the chlorophyll depletes and the carotene and xanthophyll are exposed, the color of the leaf changes.

That covers the oranges and yellows of fall trees, but what explains the fiery reds of maples that so delight tourists in eastern North America? The red in leaves can be attributed to a group of chemicals called anthocyanins, which, depending on their acidity, create a range of color from red to purple. Anthocyanins are the same molecules that give beets, grapes, and plums their hue. The mystery is: What are anthocyanins doing there?

In the sunny days of summer, anthocyanins aren't present in leaves. But when the leaves' adjustment to winter begins, the tree manufactures them. Why on earth, at a time of depleted resources, would trees go to the trouble and expense of synthesizing new chemicals? At least the others, the carotene and xanthophyll, were there already. That is the mystery of the anthocyanins, and it has generated some interesting conjectures.

One is that the anthocyanin pigments protect against sunlight. At first, that sounds completely counterintuitive. After all, gathering sunlight is what leaves are all about, and on top of that, the change in color is happening at a time of year when sunlight is in decline. But there are circumstances when too much light can overwhelm the photosynthetic apparatus and actually damage a tree's leaves. This is more likely to happen during times of inadequate nutrition and low temperatures—such as autumn—when leaves are already dismantling their internal machinery. One unusually bright day in fall could cause significant damage to a tree, but thanks to the anthocyanins in its leaves, the tree can absorb high levels of sunlight.

Some circumstantial evidence supports this idea. For one, sun-facing sides of leaves are usually redder. The leaves on the south-facing side of the tree are redder, too. And the farther north you travel, the redder the maples are—the higher the latitude, the cooler the nights, and so the more stressed the trees. If these observations are true, it suggests that trees do an incredibly delicate balancing act between getting enough light and getting too much.

The other suggestion as to why trees produce anthocyanins has attracted much more interest among scientists, likely because it lies at the heart of a fascinating theory of biological "gaming," the mechanisms and strategies by which organisms compete. Some people have proposed the idea that leaves turn red to protect themselves against insects that might lay eggs on them. For instance, in the fall, aphids lay eggs in crevices in tree bark, and these hatch after winter is over. Some scientists speculate that trees capable of manufacturing noxious, insect-inhibiting substances color their leaves red to advertise that fact to insects. Insects then lay their eggs elsewhere.

It appears aphids are in fact less attracted to red leaves than to green ones. But that doesn't mean their offspring fare worse on trees with red leaves. One experiment designed to sort this out compared wild and domesticated apple trees. Domesticated apple varieties are selected for the quality and abundance of their fruit, not for their insect resistance or leaf color. If red leaves are indeed a signal that aphids should back off, then you'd expect more red leaves in the wild varieties of apple trees, which aren't protected by farmers and pesticides. Conversely, if signaling isn't important, there should be no appreciable differences in leaf color or aphid activity between the cultivated and wild varieties.

The results of the experiment confirmed that trees with red leaves were more successful at warning aphids to stay away. The wild varieties of apple tree had a much higher percentage of red leaves in the fall than did their domestic counterparts. As well, the offspring of aphids that did lay their eggs on trees with red leaves had only about half the rate of survival of those from eggs laid on trees with yellow or green leaves.

Sadly, one good experiment does not confirm a theory, and later discoveries put this particular idea on somewhat shaky ground. It turns out that aphids have no red receptors in their eyes. But maybe they are still able to discriminate between colors. And maybe it doesn't matter: as it turns out, it might not be the red color itself that's warning insects away.

Rather, the red covers up the yellow shades of the leaf (those exposed by the removal of the green chlorophyll). Those yellows are very attractive to a broad range of insects, including aphids. So the red hue isn't a stop sign after all. But because amber means "Go" to an aphid, disguising it with red has the same effect.

What is the sound barrier and can we break it?

If you've heard an echo, then you've heard evidence that sound travels at a particular speed. The speed of sound, at sea level, is about 760 miles an hour (1,225 kilometers an hour). A fast-moving object such as a jet plane can actually catch up to the speed of sound, in effect causing that sound to pile up in front of it. When it bursts through this barrier—the sound barrier—the pressure change is heard as a sonic boom.

But planes aren't the only things that can break the sound barrier. Believe it or not, bullwhips can, too. When the tip of the cracking whip breaks the sound barrier, it creates a sudden, sharp noise.

How can a whip generate that kind of speed? It's all in the design. A bullwhip is thickest at the handle and gradually tapers toward the tip. When

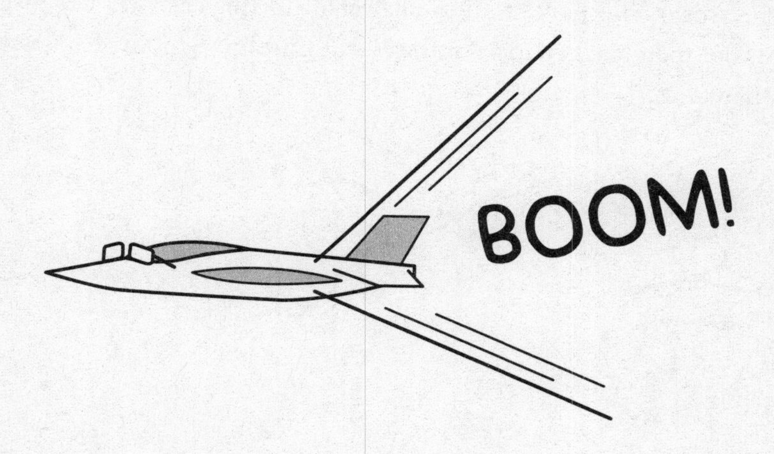

an expert cracks that whip, a wave starts from the handle and travels the length of the whip. Physics demands that the energy from the wave must be conserved. With the mass (the amount of material in the whip) gradually lessening as the wave moves toward the tip, there's more energy available to be converted into velocity, resulting in a wave that travels faster and faster. By the time the energy reaches the end of the whip, it is often going faster than the speed of sound—creating a loud sonic boom.

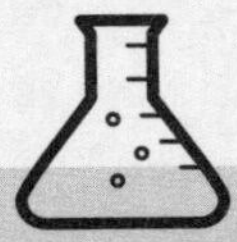

Science Fact! *One mystery that has plagued the understanding of whip cracks is that by the time you hear the crack, the tip has long passed—even doubled—the speed of sound. New evidence suggests that it is actually the loop that creates the crack rather than the tip of the whip. Either way, though, the sound is impressive.*

But bullwhips aren't the only common objects that make a cracking good sound. The poor man's version of the whip crack is the wet towel in the change room. There is no doubt that when properly flicked, the towel's sound is impressive, but is it a sonic boom? As is often the case with a totally unimportant question, there are enterprising students somewhere in the world who have tried to answer it. In this case, it's not just students but their teacher, too—at the North Carolina School of Science and Mathematics.

In 1993, they recorded up to ten thousand images per second of a flicked towel. It was a demanding setup, and they were rewarded by captured images of the tip of the towel clearly surpassing the speed of sound! To their credit, they knew this wasn't enough evidence to prove that the sound was a sonic boom, just that the towel had reached sufficient velocity to produce one. So they modified their experiment, using sound to trigger the photo-taking. They found that the highest velocity of the towel tip and the blast of sound coincided. Still, there were puzzles, because they also recorded a snap or two when the towel did not exceed the speed of sound. They reasoned that in these cases the period of time the towel broke the sound barrier was briefer than the intervals they were recording. When they reduced those intervals, they proved that indeed the sound barrier had been breached.

That was pretty cool work (and nobody appears to have replicated it since), but the absolute ultimate in flicking something to break the sound barrier takes us back something like 150,000,000 years, to the age of the dinosaurs—specifically, to the giant herbivore *Apatosaurus*. Yes, that's right: some dinosaurs might have been able to snap their tails so fast that they broke the sound barrier.

The *Apatosaurus* was huge: sixteen to seventeen tons (fifteen to sixteen tonnes)—three to four times the weight of an African elephant—and sixty-five feet (twenty meters) long. Just over half of the creature's total length was devoted to just the tail. The tail of *Apatosaurus* and a bullwhip have one thing in common: their shape. They're thick at the base and taper all the way down to the tip. In a bullwhip, the handle is about six hundred times thicker than the tip. In *Apatosaurus*, that taper was even more dramatic: from three feet (one meter) thick where it attached to the body to a tip that was thirteen hundred times smaller. Each vertebra in the tail was 6 percent smaller than the one before, and there were more than eighty of them. Amazingly, the first twenty or so vertebrae supported 97 percent of the total mass of the tail.

Some have suggested that the very long tail was a defensive weapon. Others posit that it might have been part of a mating display. Nathan Myhrvold, former chief technology officer at Microsoft, and Phil Currie, a paleontologist at the University of Alberta, were intrigued enough by the tail's peculiar size and shape to investigate further. They wondered: With a tail that unique and that long, if the *Apatosaurus* flicked it, would it create a bullwhip cracking sound?

Working with one of the best *Apatosaurus* skeletons and filling in details from other skeletons where necessary, they reconstructed the length, weight, and flexibility of the tail. They concluded that it was indeed enough like a bullwhip to have generated a fast-moving wave along its length. There were other encouraging details, too, like the fact that the vertebrae toward the tip had gaps between them, suggesting the presence of cartilage padding that would help with the flicking. The tail also seemed to lash horizontally rather than vertically, which made sense given the design of the vertebrae and the supporting structures around them.

If the animal whipped the base of its tail sideways at a speed of something like 6½ feet (2 meters) per second, that would magnify to propel the tip over the speed of sound. And it would require less energy than walking! Given that the animal carried 70 percent of its weight in its hindquarters, simply stamping its feet could send the tail into a lash. So what would a dinosaur tail breaking the sound barrier sound like? Myhrvold and Currie calculated a noise that might have reached two hundred decibels—enough to startle even a *T. rex*!

There were critics of Myhrvold and Currie's theory, of course, and in response Myhrvold built a replica of the hindquarters and tail of the *Apatosaurus*. It is quarter-sized, weighing 44 pounds (20 kilograms) with a length of merely 11 feet (3.5 meters), but it is exactly scaled to the tail of the dinosaur, complete with eighty-two vertebrae constructed of steel, neoprene, aluminum, and Teflon. When you crank the tail, the tip breaks the sound barrier and creates that satisfying crack, suggesting that it may well have been possible that this dino broke the sound barrier 150,000,000 years ago. So it's not a tall tale about a long tail, but rather as Currie and Myhrvold called it: supersonic sauropods.

History Mystery

Is it true that right now we are breathing the same air that Julius Caesar breathed?

That little word "air" causes confusion. Air isn't a single thing; it's a mix of several gases. Nitrogen and oxygen make up 99 percent of the total, along with whiffs of argon, carbon dioxide, water vapor, and other trace gases. Every time Caesar exhaled, molecules floated off in all directions, dispersed by the wind. But could any of the atoms contained in one of Caesar's breaths from more than 2,060 years ago still be around for you to inhale? The answer is yes!

Why is that? Atmospheric gases such as nitrogen, oxygen, and carbon dioxide are made up of molecules, which are in turn formed from clusters of atoms. Over time, all molecules are subject to chemical and radiative forces that break them apart, but the atoms themselves aren't destroyed. Two single oxygen atoms that combine to make up an oxygen molecule might split up, then go on to attach to different oxygen atoms or perhaps latch on to a carbon atom to make carbon dioxide. It's an endless cycle. The exact molecules that Caesar breathed aren't likely to be floating around anymore, but the atoms that composed them are just waiting for you and your lungs.

Did You Know . . . Julius Caesar was assassinated on March 15, or the ides of March, in 44 BCE, a day that became notorious because of his murder. The air in his last breath out had about 4 percent less oxygen and 4 percent more carbon dioxide than it did when he inhaled it, because oxygen is used as fuel by our bodies, with carbon dioxide being the resulting waste. Caesar was stabbed twenty-three times by Brutus and dozens of other Roman senators, although apparently only one wound was fatal. More like a last gasp, then, not a breath.

Assuming Caesar was of average size, the amount of air he exhaled in a single breath would have been 30.5 to 61 cubic inches (about half a liter to 1 liter). A well-known formula tells us the number of molecules contained in that volume at an air temperature of 32 degrees Fahrenheit (zero degrees Celsius). Even accounting for Rome's slightly warmer, less dense air in March, a

fair estimate is 1 x 10^{22}—or 10,000,000,000,000,000,000,000 molecules in just one exhalation.

Because more than two thousand years have passed since Caesar's death, we can assume that the air he breathed has been thoroughly distributed throughout the atmosphere. And, as you'd expect, there is an astounding number of molecules in the earth's atmosphere: 1.08 followed by 44 zeroes.

Okay, so Caesar exhaled 1 x 10^{22} molecules; let's say you inhale about the same, although if you take a really, really deep breath you might be able to take in twice as many—but with these numbers, twice as many makes almost no difference.

Those molecules were then diluted into a volume of air that was—and is—gazillions of times bigger. (A gazillion is not a real number, by the way.) On the other hand, all you need is one molecule of Caesar's to enter your lungs. Now it's simple division:

> 1.08 x 10^{44} (molecules in the atmosphere) divided by 1 x 10^{22} (molecules in one breath) = 1 x 10^{22}

So you'd need to inhale (roughly) 1 x 10^{22} molecules of air to get one of Caesar's, and that is exactly what you're doing! So it's true: with every breath you take, there's a very good chance you're inhaling one of the molecules (or, to be pedantically correct, at least one of the atoms) that was part of Caesar's last breath. Keep breathing . . . big, deep breaths . . . and you are creating an even stronger bond with the great Roman leader.

Did You Know . . . Over the course of the average lifetime, a human being breathes around 500,000,000 to 700,000,000 times, taking about 900 breaths per hour. In comparison, the average dog breathes less than 200,000,000 times in its life—even though it's breathing twice as fast as a human, it lives a much shorter life.

Part 5
Weird Science & Machines

Could we bring back the dinosaurs?

There's a reason that *Jurassic Park* was a movie and not a scientific project: making movies is easier than bringing back a dinosaur—and, even with mega-budgets, cheaper. And bringing back dinosaurs might not be the smartest thing to do.

The first challenge is to recover pristine dinosaur DNA. In *Jurassic Park* that DNA came from a mosquito filled with dinosaur blood that then blundered into still-liquid tree sap. That sap hardened, eventually becoming amber, handily preserving the insect for more than 60 million years.

How likely is that scenario? The closest we've come is the discovery of the remains of a 46-million-year-old blood-filled mosquito preserved in Montana shale. That's nearly 20 million years after the dinosaurs died out, so not much help in re-creating *T. rex*. There are only two other

known mosquito fossils old enough to have coexisted with the dinosaurs: one, from Burma, contains material that hasn't yet been analyzed; the other, from Alberta, is a male and therefore wouldn't have bitten anything. No bite, no blood; no blood, no dino DNA!

That doesn't mean we won't find the perfect specimen one day. The fact that blood is preserved in the 46-million-year-old specimen is amazing and hopeful: even the hard shells of beetles degrade over that much time. But it will likely be a flea-like insect instead. Getting our hands on the blood still leaves us far short of *Jurassic Park*, but there is good news: there's now evidence that we may not even have to bother with the insect.

Over the last decade several labs have found dinosaur tissue in the form of proteins preserved in dinosaur fossils themselves, a discovery that came as a huge surprise, given that fossils are, by their very nature, rock. Now, proteins are one thing, DNA is another, but the fact that there is even this much preservation suggests we're not at a dead end yet.

Let's fantasize that preserved DNA is found, it's actually dino DNA, it's intact, it can be extracted and the amounts can even be amplified so they can be worked with. Then what? Apply the techniques of cloning!

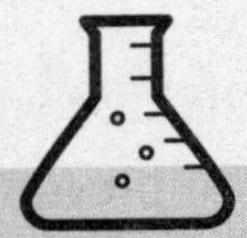

Science Fact! *Cloning has been very successful with modern mammals—remember Dolly the sheep?—but those mammals are the ideal. In Dolly's case it was easy to get a living sheep's DNA, place it into the nucleus of a sheep egg, transplant that egg into a female sheep, and let pregnancy take over. But Dolly was only one success in 277 attempts!*

When it comes to dinosaur cloning, we have no DNA, no viable eggs (the ones we have are fossils), and no female. There are possible solutions. The closest living relatives of the dinosaurs are birds, so you could implant dino DNA into an ostrich egg and in turn transfer the egg into a female ostrich. (Fertilization isn't needed because the DNA already has the genetic contribution from both parents.)

If all you have on hand is fragmentary dino DNA, you could add those dinosaur genes into a set of ostrich genes. That would surely be a trade-off, because the offspring would end up more ostrich than dino, but at least you'd enhance its chances of survival. Even with a complete set of dinosaur genes, the development of the embryo would be guided by sets of carefully timed inputs from the mother, so the offspring would at best be some sort of weird bird-dino hybrid. A single weird bird-dino hybrid if it indeed survived.

And that single offspring would have to eat. Birds are the descendants of therapod dinosaurs—like *T. rex* or velociraptor—who were carnivores. But a healthy existence for such creatures includes more than just meat on the hoof: gut bacteria are essential. Where would we find those? Finally, to set up a self-sustaining population, at least five thousand more animals would have to be produced and housed on a piece of land at least the area of a national park. Our ignorance of dinosaurs' ecological needs would almost certainly doom such a vast project.

If we're willing to settle for something less than a dinosaur, however, the picture gets brighter. Take the woolly mammoth. There are frozen specimens less than five thousand years old, and high-class mammoth DNA has been recovered; in fact, the complete mammoth genome has been sequenced. Second, the modern elephant, in particular the Indian species, is closely related. Genetic techniques could be used to substitute mammoth genes for Indian elephant genes, then the mammoths could be cloned.

Geneticist George Church at Harvard has already spliced mammoth genes into Indian elephant DNA—forty-plus so far. He's selected the most relevant genes for survival in cold climates, including smaller ears, hair, layers of fat, and even blood that's efficient at transporting oxygen in cold climates. That's fantastic progress, but there are still huge hurdles to be overcome.

For one thing, those forty-plus genes are only a small fraction of the genetic differences between mammoths and modern elephants. Not to mention the fact that the elephant and mammoth genes might not play nice together. For another, Church has said that he'd raise the embryo in an artificial womb: he couldn't justify experimenting with the Indian elephant, which is endangered. But that artificial womb would need to shelter an animal that could take twenty-two months to reach maturity and grow to more than 200 pounds (90.7 kilograms).

The mammoth genome chosen to represent the species is itself a challenge. It was derived from the last surviving population living on Wrangel Island in Arctic Russia, and that population had developed serious genetic flaws through inbreeding that likely contributed to its extinction.

The Russians are developing something called "Pleistocene Park" in Siberia, but even if Church is successful, there's no guarantee that reborn mammoths would have a place to live. And that's a problem that would haunt any species we bring back.

One final thought: it's likely that the money and attention given to these charismatic animals of the past—the mammoth, the dodo, and the passenger pigeon—would be better spent on saving the countless species that are still alive today but are threatened with extinction.

Are we alone in the universe or are aliens out there?

You can't even start to answer this question unless you believe that life—intelligent life at that—could have begun on planets other than Earth. You don't have to believe that: it's still quite possible that we are unique in the universe—that no matter how many billions of galaxies exist, containing billions of stars that have untold billions of planets orbiting them, we're the only ones. But the attitude that we're the center of everything has been eroding since the 1500s and has reduced us from being the one and only to being one of eight planets orbiting a humdrum star in one of an incalculable number of galaxies.

It's challenging to figure out whether we're alone in the universe when we don't yet have evidence of life anywhere else. But there's a way of

approaching it, mostly thanks to astronomer Frank Drake, who, in 1961, invented something called the Drake equation.

The Drake equation is a series of unknown quantities that give a sense of what we have to know before we can be confident that there are other intelligent civilizations out there. It's written like this:

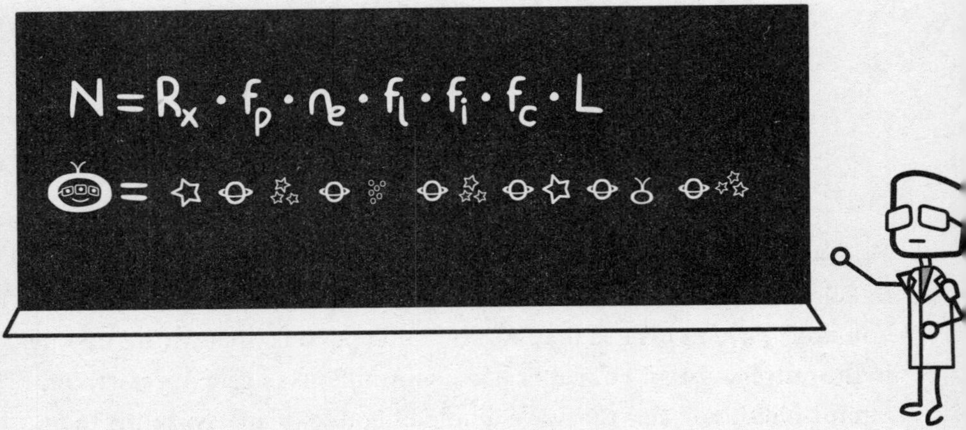

$$N = R_x \cdot f_p \cdot n_e \cdot f_l \cdot f_i \cdot f_c \cdot L$$

Translated into English, the equation says that *N* is the number of technologically advanced civilizations out there right now that we might be able to discover. Exciting stuff! *N* means extraterrestrials. *N* means aliens!

But *N* is dependent on everything to the right of the equals sign. As each term is taken into account, *N* shrinks. That means the chances of us finding another species in the universe, then, is based on:

R_x = the total number of stars
f_p = the fraction of those stars with planets
n_e = the number of planets that are the right distance from their star to allow the existence of life
f_l = those planets that actually do support life
f_i = those where intelligent life managed to evolve
f_c = the ones that acquired advanced communications technology and the last number,

L = the number of technological civilizations that actually survive long enough for us to detect them.

When Drake came up with his equation, many of the numbers in it could only be guessed at. But since then we've managed to get a little more exact.

Science Fiction! *We are most familiar with life-forms evolving on land or in water. But are there other possibilities? Two famous astronomers, Fred Hoyle and Carl Sagan, imagined weird and wild gaseous life-forms. In Hoyle's late-1950s science fiction novel* The Black Cloud, *a giant cloud of dust and gas invades our solar system and blocks out the sun, threatening all life on earth. The cloud, more intelligent than us, lived off the energy of radiation from stars—what we call sunlight. Earth was saved when the cloud decided to move on. Carl Sagan, in a paper for NASA, put forth the idea of three kinds of giant balloon-like organisms existing in the atmosphere of Jupiter: floaters, sinkers, and predators. Floaters would be kilometers in size and would survive by gathering sunlight or processing the chemicals in the atmosphere. Sinkers, like the ocean's plankton, would slowly fall through the atmosphere but could absorb other things as they fell (such as floaters), the way raindrops grow as they fall. And hunters, of course, would target other organisms to absorb.*

Planets orbit stars, so, to start, we need to know how many stars there are in the universe. Our galaxy, the Milky Way, has at least 100 billion stars, and that could be roughly the same number in any galaxy. There are somewhere between 10 billion and 10 trillion galaxies, so if you multiply those numbers (using the larger estimate of galaxies), you get an incomprehensible 1,000,000,000,000,000,000,000,000. Lots of stars.

Did You Know . . . There are different estimates, but there could be as many as 60 billion habitable planets in the Milky Way galaxy.

New technologies, like the Kepler space observatory, have given us a much better idea of how many stars actually have planets around them. There are many planets out there, but we expect the ones that might hold life are those that are roughly the same size as Earth and located in what's called the "habitable" zone, where water can exist as a liquid. We earthlings assume that water, crucial to life on Earth, would be equally important elsewhere. That means that a planet can't be too close to its sun (where the heat would evaporate the water) or too distant and cold (where the water would freeze).

We have already discovered more than 4,000 planets orbiting other stars, and it's likely that, on average, every star has at least one planet, and at least one star out of every five has an Earth-sized planet in its habitable zone. And that's not including the claim that more than 90 percent of the galaxy's planets have yet to be created. A planet's size is important, too, as it's harder for life to evolve on a giant gas planet like Saturn than it is on a rocky planet like ours.

Unfortunately, we have no idea how likely it is that a planet—even one in the habitable zone—can support life. So far we only have one example—us—in our solar system. That makes it hard to guess about elsewhere, but even if evidence of past microbial life were found on Mars, that would change the odds considerably. Scientists are hopeful that life might be common, because the chemical compounds crucial for life aren't limited to Earth at all but are found scattered all over the galaxy.

As difficult as it is to estimate how widespread life might be, what about intelligent life? Here, although it's really a guess, scientists seem comfortable with the idea that if you find ten planets with life on them, it's likely that one will have intelligent life. What's much more important is whether those intelligent species are able to become technologically adept, because only then will we be able to detect or even communicate with them.

Did You Know . . . Philosopher Nick Bostrom has argued that we don't want to find other species in the universe. According to Bostrom, the rarity of intelligent life in the universe is proof that there is some event, a crucial barrier, that holds back all but a very few lucky technological civilizations, and that so far we're the only example of that.

Why is this important? If this crucial step is in our past, we're successfully through it, and the fact that we seem to be the only ones to have made it suggests that achieving technology is a very rare event. But if that barrier to becoming a fully technological, space-exploring civilization is still ahead of us—if many planets have already reached the stage we're at now, and moved on, why do we see no evidence of them?

Funnily enough, what we find on Mars is important to Bostrom's theory. He is hoping we won't find a single trace of microbial life on Mars, because that would signal that life happens often on other planets. And if that's true, it's much more likely that intelligent life, like ours, has appeared elsewhere and has been wiped out. For Bostrom, if there's no life on Mars, we can dream that we're unique. But if there is, that might suggest a bad future for our species.

That brings us to the last two numbers in the Drake equation. Detecting technologically advanced species would be awesome, but communication with them is the real goal. We have been a technological species for at the very most a few million years. (Stone tools 3.3 million years old have been found in Kenya.) And technology allowing us to communicate with distant civilizations has been around for only about a hundred years. That's not very long when you consider the lifetime of the planet—4.6 billion years—and it isn't very much time for another civilization to find us. With that timetable, aliens could have been calling us for millennia and given up long ago because we didn't answer!

Did You Know . . . We have been inadvertently sending signals to extraterrestrials for longer than we've been listening for them. Before cable, TV signals used to be literally broadcast through the air. Those signals may well have been traveling through space. Just think: programs like Rod Serling's *Twilight Zone* have been traveling at near light speed since 1959, putting it somewhere between 50 and 60 light-years out there. (Would *The Twilight Zone* freak out aliens?) Unfortunately, most radio broadcasts never make it outside earth's atmosphere.

It's clear that, despite all of our advancements, we still don't have exact numbers for all of the terms in Drake's equation, which makes it impossible to come to a conclusion about alien life. Solutions to the equation range from one civilization in our galaxy (us) possibly hosting technologically advanced life to hundreds if not thousands.

Scientists have now started varying the Drake equation to ask: How likely is it that an intelligent civilization has ever arisen in the universe? The conclusion was that unless the odds are worse than 1 in 10 billion trillion (1 in 10,000,000,000,000,000,000,000), intelligent life has to have happened. Surely the odds have to be better than that, right? Of course, we haven't heard from any of these civilizations yet, but we keep hoping they'll make contact. Maybe they're just waiting for an invitation.

Could we ever build a space elevator?

What is the highest elevator you've ever been in? Sixty stories? Seventy? One hundred? How about an elevator that is 12 million stories? Even on an express elevator, that would still be one long ride. But that's what it would take to create an elevator that runs from the earth's surface to the upper limits of our atmosphere—to space.

Building a space elevator isn't impossible, it's just really challenging—the equivalent to building a suspension bridge around the world. The first problem is that the elevator shaft can't be built from the ground up. There simply isn't a construction material in existence that could support

something that big. You'd need a base the size of a mountain, and even with that, the whole thing would collapse under its own weight long before it reached orbit.

But there's good news! Rather than starting from the bottom, you could assemble the elevator shaft from a satellite orbiting earth and lower it through the planet's atmosphere as it grows. Physicists love this approach: as it's being built, the tower would experience tension, not compression, and that would help prevent its collapse.

Did You Know . . . The notion of a space elevator began with the idea of so-called geostationary satellites. In 1945, science fiction author Arthur C. Clarke pointed out that if you launch a satellite into an orbit at exactly the right altitude above the equator (just under 22,369 miles, or 36,000 kilometers), it will whip around the earth at the same speed that the earth is rotating and will appear not to be moving at all when viewed from the ground.

As brilliant an idea as that was, Clarke didn't take it any further. But a Russian scientist, Yuri Artsutanov, did. Artsutanov asked: Why, if a satellite can hover over the same place on earth all the time, can't two be connected? And if an elevator shaft joined the two, then it would be possible to take an elevator to space instead of a rocket. (Although more recently the space elevator has been envisioned as a simple cable with the equivalent of gondolas running up and down.)

There is, however, one small but crucial detail standing in the way of this breakthrough. As the shaft is assembled downward, the gravitational force it feels will increase the closer it gets to the earth, dragging the satellite out of orbit. To stop that, a cable of the same mass would have to be extended above the satellite into space, exerting an equivalent force upward to counterbalance the pull of the elevator toward the earth.

This two-pronged approach would reduce some of the dangerous risks considerably, but it would bring its own challenges. For instance, by the time the elevator shaft/cable reached the earth, the outward, stabilizing extension, if it were about the same diameter, would have to be more than 62,137 miles (100,000 kilometers) long. The entire setup, upward- and downward-extending cables together, would extend about a third of the distance to the moon. A suggested solution—if it's actually fair to call it that—is to capture an asteroid and tow it into place just above the elevator, providing the necessary mass in a somewhat more compact package.

When the idea was first discussed in the 1970s, there was no material strong enough to build anything as big as a space elevator. The best suggestions were special forms of carbon, like crystals or graphite, but those were still inadequate. Luckily, the science of strong materials has come

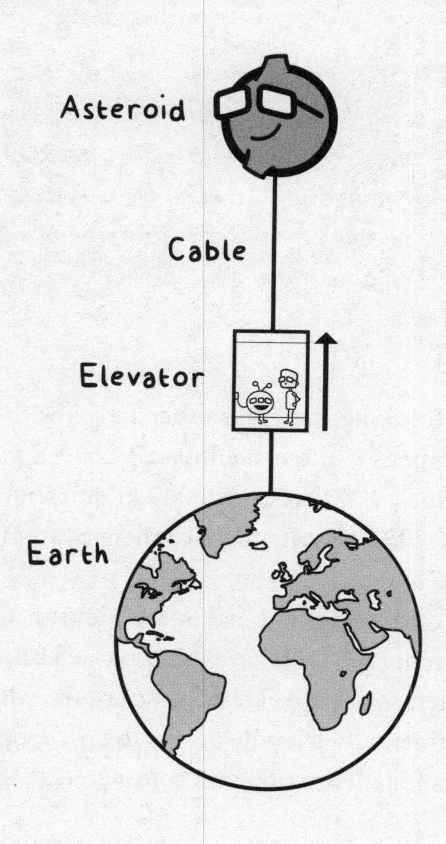

a long way since then, and nanotechnology has completely changed the picture.

The best material currently available is a peculiar form of carbon called *buckminsterfullerene*, named after inventor Buckminster Fuller because the molecules can form micro-geodesic domes that resemble the ones he designed. But they also form tubes called "buckytubes." Buckytubes are incredible: they're resistant to breakage and tension (the main force that a space elevator has to fight), and a rope of buckytubes less than an inch thick would be a hundred times stronger than the same rope made of steel while weighing only one-sixth as much. But ropes of buckytubes have yet to be built. Other experimental materials are out there, but they're also a long way away from being ready for use.

Science Fiction! *After coming up with the idea of satellites in geostationary orbit, Arthur C. Clarke proposed linking a number together into a "ring city" complete with a railroad. And in 1951 Buckminster Fuller suggested building a "halo-bridge" above the equator to which people could climb from one place on the earth, ride for a while, then descend at another.*

You might be thinking there are other issues with building a space elevator. What happens if the cable breaks? Or if a plane collides with it? Or if you get stuck 9,321 miles (15,000 kilometers) up with a bunch of people you don't like? Despite these challenges and more, though, the idea lives on.

The president of the International Space Elevator Consortium, Peter Swan, released a report in 2017 in which he defended the reasons for building a space elevator. Space elevators, he argues, will offer cheap, safe, and more environmentally friendly access to space, opening up unforeseen opportunities for exploration and commerce.

Swan's vision of the elevator is a meter-wide ribbon on which small electric wheeled vehicles would climb, clinging to the ribbon by friction. He admits that it will be at least another ten years before the right carbon nanotubes or their equivalent will be available to construct the ribbon, but he remains not just optimistic but even poetic, comparing the thunderous takeoff of a rocket to the noise created by the ascending elevator as being like "dropping petals into a pond."

The space elevator might simply be an exercise in imagination—or it might turn out to be something far beyond that. But all of this fiction and possible fact shows our fascination with the idea that we might someday all have a chance to press our noses against the glass and stare down at the earth from 22,369 miles (36,000 kilometers) above.

What's dangerous about the Bermuda Triangle?

The Bermuda Triangle has a bad reputation. A huge expanse covering more than 500,000 square kilometers of ocean, it is associated with more than a hundred unexplained disasters. The points of the triangle are anchored by Florida, Puerto Rico, and, of course, Bermuda.

So is there something truly dangerous about the Bermuda Triangle? And, if so, what is it? Some theories argue that "crystal energy" accounts

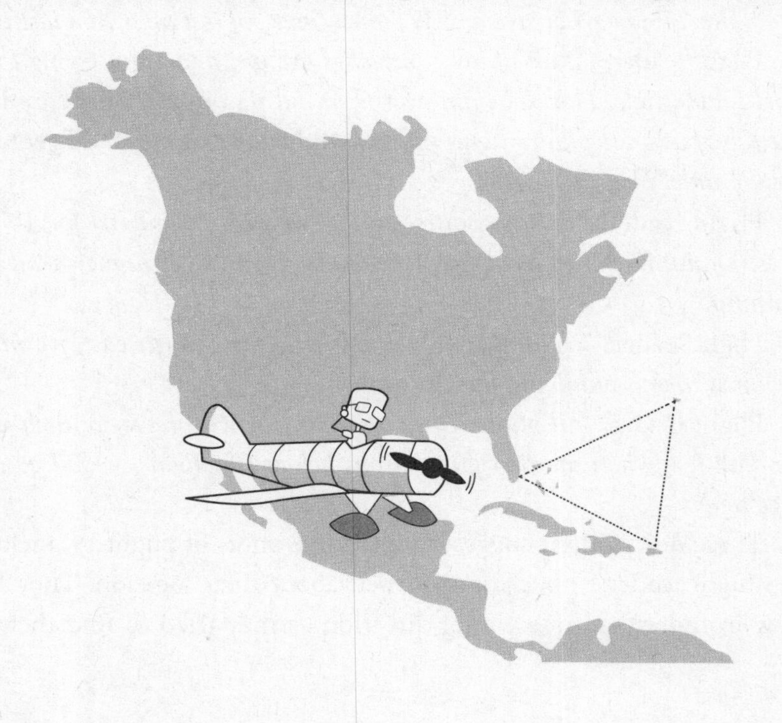

for the strange disappearance of planes and boats in the triangle. Others say that alien interference in the region is to blame. Still others point the finger at spirits. Or a giant squid. Or an interdimensional doorway. Before you go too far down the supernatural spiral (taking your spirits and giant squids with you), let's check out the scientific evidence.

Accounts of ships and planes disappearing in the Triangle have been documented for well over a hundred years. The most famous of these disappearances occurred in 1945 when five TBM Avenger torpedo bombers with the U.S. Air Force (later named the Lost Patrol) disappeared. The planes took off from Fort Lauderdale just after 2 p.m. on December 5 on a routine practice run. The plan was to fly due east over the Atlantic Ocean to engage in a brief practice bombing, then fly farther east, then north, and finally west again to return to base. The weather was excellent when the fleet took off, but the planes never made it back to base. No trace of any of the aircraft was ever found—no wreckage, no bodies. Fourteen lives were lost, and the mystery remains unsolved.

The only evidence we have are the radio reports from the pilots and they suggest the pilots were confused as to where they were:

"I don't know where we are. We must have got lost after that last turn."

Flight leader: *"Both of my compasses are out and I am trying to find Fort Lauderdale, Florida. I am over land but it's broken. I am sure I'm in the Keys but I don't know how far down and I don't know how to get to Fort Lauderdale."*

Flight leader: *"Change course to 090 degrees [due east] for 10 minutes. Dammit, if we could just fly west we would get home; head west, dammit."*

Flight leader: *"Holding 270, we didn't fly far enough east, we may as well just turn around and fly east again."*

Flight leader: *"All planes close up tight . . . we'll have to ditch unless landfall . . . when the first plane drops below 10 gallons, we all go down together."*

These descriptions suggest that the five pilots of Flight 19, including the flight leader(!), became confused about their location. They likely flew in more than one wrong direction as they tried to find their way back to Florida, finally running out of fuel and crashing into the ocean. True, their compasses had stopped working and the weather turned stormy. But still, there is no clear explanation for why all five would have gotten so lost.

The disappearance of those five planes was strange enough, but another thirteen people died right after when a different aircraft—one of a pair sent out to search for the missing planes—was also lost. The rescue plane, a PBM-5, radioed back to base three minutes after takeoff, and that was the last it was heard from. Twenty minutes later a ship in the area, the SS *Gaines Mills*, reported seeing an explosion and "flames 100 feet high." Shortly after, the ship sailed through a slick of oil and aircraft fuel. The downed rescue plane was reputed to have had recurring problems with fuel leaks. Even with precise details about the location of the explosion, not a single piece of that aircraft was ever found.

Science Fiction! *The Bermuda Triangle is the final resting place for the city of Atlantis . . . or so say Paul Weinzweig and Pauline Zalitzki. The two scientists claim to have captured images of sphinxes and pyramids of the Lost City on the ocean floor in the Bermuda Triangle. There's just one problem: given the location of these findings, Atlantis could only ever have existed hundreds of meters below sea level, which would have made it a very wet city. Weinzweig has acknowledged that the pyramidal shapes he found on the ocean floor might be natural formations, not from Atlantis at all.*

The USS *Cyclops*, a ship that disappeared without trace in 1918 with more than three hundred people on board, is another well-known disappearance in the Triangle. The ship was likely vulnerable to stormy weather because it was carrying about eleven thousand tons of manganese ore, making it tons over capacity and seriously overweight. Two of its sister ships later sank because of suspected structural flaws.

The latest attempt to explain the supposed "weirdness" of the many disappearances in the Bermuda Triangle brings more science into it. The Science Channel in the USA promoted a theory that unusual hexagonal cloud formations seen in satellite images over the Triangle might be connected to the disasters there. According to the report, the hexagonal clouds generate microbursts, downward "air bombs" that travel at speeds of up to 186 miles (300 kilometers) per hour. These would obviously be hazardous to boats and planes, but so far this theory has not been proven.

The bottom line is that while there are no definitive explanations for the disappearances in the Bermuda Triangle, in most cases, there's some rational evidence to suggest mechanical failures and weather events as opposed to squids, crystals, and spirits. In the end, it seems that the most dangerous thing about the Bermuda Triangle is simply believing in it.

Does Bigfoot exist?

Most of the subjects in this book qualify as science, but sometimes I like to write about "science." "Science" includes topics that have a scientific veneer, but a little scraping and sanding reveals there's nothing much underneath. UFOs, spontaneous human combustion, and the Loch Ness monster are all perfect examples. Some of these "science" topics have quietly vacated the headlines, but others linger, no matter how little evidence supports their existence. I am fascinated—not because there's evidence validating these topics but because of people who labor long and hard to maintain their belief in them despite that lack of real evidence. One of my favorite "science" topics is that great man-ape of the Pacific Northwest, Bigfoot, or the name I prefer, Sasquatch.

It would require several books to give proper due to the Sasquatch legend, but I'll do my best here to probe into the "evidence" of its existence, knowing as I do that many believers will think I've shortchanged the animal terribly.

After all these years, the prime piece of evidence remains the Patterson-Gimlin film of 1967. That year, Bob Gimlin and Roger Patterson were drawn to Bluff Creek, California, by rumors of mysterious large footprints found in the forests. The pair took their camera with them in the hopes of spotting the creature that had made the prints. And lo and behold, they did!

The film they shot depicts a large human-like, two-legged furry animal walking unhurriedly from left to right across a clearing in a forest. The footage is pretty shaky. Patterson claims his horse reared at the appearance or scent of the creature and threw him. Only when he regained his equilibrium was he able to aim the camera properly and get a better shot of the beast.

A skeptic's first thought at seeing the film is that this was a guy in a gorilla suit, albeit a pretty convincing one. Fortunately, I was lucky enough to attend a two-day Sasquatch conference at the University of British Columbia Museum of Anthropology in 1978, where the gorilla-suit claim and other Sasquatch-related clues were examined closely. At the conference, Russian and North American Sasquatch acolytes dissected, slowed, paused, rewound, reshowed, worshipped, and critiqued the Patterson-Gimlin film, all in an effort to come up with definitive evidence that this was some new creature, not a human in disguise.

Some said the figure was too broad-chested to be a human; subsequent research has shown that not to be true. One expert claimed that if the film had been shot at twenty-four frames per second, the thing was walking like a human, but if it had been shot at sixteen frames per second, it wasn't. Patterson wasn't sure whether the camera had been set at sixteen or twenty-four frames per second. How handy for the perpetuation of the story that there was this uncertainty!

Next on the docket: an examination of the movements of muscles underneath the skin of the creature. It was argued that rippling muscles would be impossible for a man wearing a loose-fitting gorilla suit. The rotation of the torso as the animal glances back at the camera couldn't be accomplished believably with a suit, and the arms were too long to be a human's. Much discussion was given to the fact that the creature had "pendulous breasts" and therefore must be a female.

John Napier, an orthopedic surgeon, paleoanthropologist, and solid science type, weighed in on the matter. A bona fide expert on foot structure in human beings and apes, he noted that the animal walked like a man, but with a strange cadence and exaggerated arm movements. Then he got specific, pointing out that the skull of the beast had a crest on top, like a male gorilla's, but unlike that of any female primate. Could this suggest that someone with not quite enough knowledge of apes and humans had designed this animal?

Napier wasn't finished. He argued that a Sasquatch's center of gravity would be higher than a human being's, and that would change its gait. But this animal walked like a human, and must therefore have a human-like center of gravity. If it looks like a duck and walks like a duck . . .

But it was the creature's buttocks that sealed the deal. Napier noted they didn't belong—at least they didn't match the upper body. Muscular buttocks are a human feature, so while Napier thought the upper half of the creature was ape-like, the lower half was human-like. That's impossible naturally, so one of those halves had to be artificial. He was pretty sure it was the upper half. But even as he raised doubt, Napier refused to shut the door totally on Sasquatch. He felt, partly based on the number of people who claimed to have seen it, that it was still impossible to say, "It does not exist."

Greg Long, however, was driven to put the Sasquatch story to rest forever. In 2004, he published a book called *The Making of Bigfoot: The Inside Story*, in which he claimed to have found both the man who wore the gorilla suit and the man who supplied the suit. He identified the suit wearer as Bob Hieronimus. Long claims to have taped Hieronimus both walking normally and walking as a Bigfoot and put that tape side by side with the Patterson film. The two matched. Admittedly, there are inconsistencies in Long's account: Hieronimus remembers the suit being a three-piece affair, while the man who purportedly made it claims it had six pieces. Why the breasts? Apparently, they didn't come with the suit. Why did Hieronimus remember the suit having a terrible smell when it

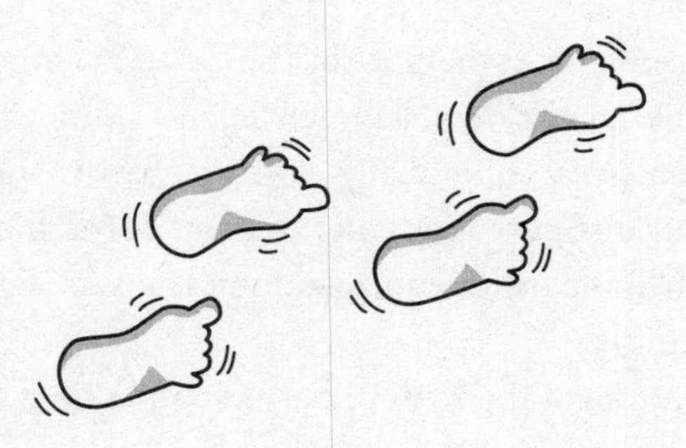

was an off-the-shelf gorilla suit? Why do Hieronimus's legs and arms not correlate with those of the animal on film? And what happened to the suit? All good questions with no answers.

But the film footage of Sasquatch is only one piece of so-called evidence. What about the footprints? Sasquatch footprints are scattered across the Pacific Northwest. Over a hundred thousand have been found in the past fifty years, most of them obviously faked but a fraction that are harder to explain away.

I'll take an idealized print, one of the more believable samples, and start there. First, there's no arch. There is no arch because the tendons and ligaments that maintain the arch would be unable to support the weight of this animal. Second, the big toe is no larger than the rest. We humans stride off our big toe, but we don't weigh as much as a Sasquatch. And third, the heel leaves a deep imprint on the inside of the foot, not the outside, a reflection of the animal's weight. All these differences could be consistent with a nonhuman, ape-like form.

John Napier was convinced there were two kinds of Sasquatch footprints, implying two different species. He declared that one had to be fake. At the same time, he maintained there was "a curious and persuasive consistency" about the footprints, especially the variety he considered natural: the so-called Bossburg prints.

Found in 1969 near Bossburg, Washington, these prints formed a trail of 1,089 impressions. They were huge—17 ½ inches (45 centimeters) long and 7 inches (18 centimeters) wide—but more significant, this Sasquatch had a clubfoot. One footprint was normal, but the other curled into the shape of a parenthesis, with what looked like calluses on the outside of the foot.

Experts argued about whether this reflected an inherited condition or was caused by an accident. Human beings born with this congenital *talipes equinovarus*, colloquially called clubfoot, often rest only the front part of their disfigured foot on the ground. But these prints clearly showed the heel. Some felt the unusual footprint was therefore the result of a childhood injury.

Are you as far down the rabbit hole as I am? These are anatomically correct clubfooted Sasquatch prints. Who would fake those? Even better, who would be qualified to fake those? Sasquatch believers claim a faker with such skills would have to be the equivalent of an artist/scientist like Leonardo da Vinci, thereby proving the beast is real. Sasquatch skeptics point out that books full of illustrations of disfigured feet exactly matching the Sasquatch print can be found at the Washington State University libraries at Spokane, two hours away.

If there's no Sasquatch—and I don't see the chances getting any better—then every footprint is a fake. And that means that many individuals from Alaska to California have, directly or indirectly, well or poorly, conspired to create a mass footprinting of the land. It's hard to imagine, but it's even harder to think the creature has eluded us all these years.

There are always those who pursue and perpetuate supernatural mysteries with incredible fervor—they are often more fascinating than the mysteries themselves. For the record, I lost interest in Sasquatch when one was spotted in a mall in Wisconsin.

To pee or not to pee?

How does one pick the most private urinal in a public bathroom?

Many men's public bathrooms still host a wall of urinals. There's a common belief that when men are at urinals, if they can, they will avoid standing beside another man. A space in between is so much more comfortable. So a man walks into a bathroom and is faced with a choice: Which urinal is most likely to have an open stall on both sides? How might he minimize the chances that someone will move in next to him?

A study has been done on this, believe it or not. In their paper called "The Urinal Problem," Evangelos Kranakis and Danny Krizanc take a math and computing science approach to the issue.

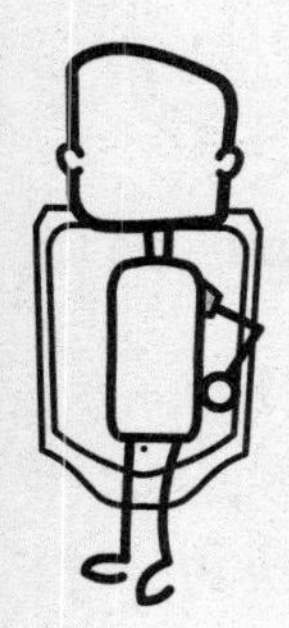
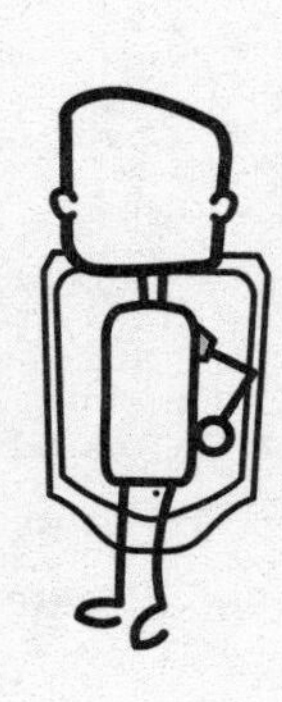
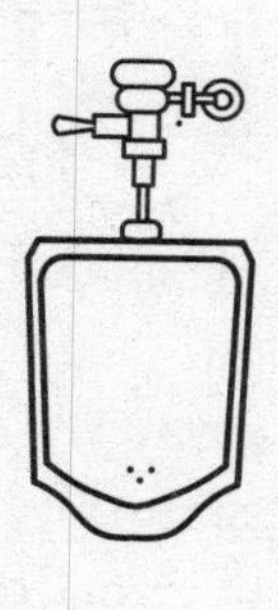
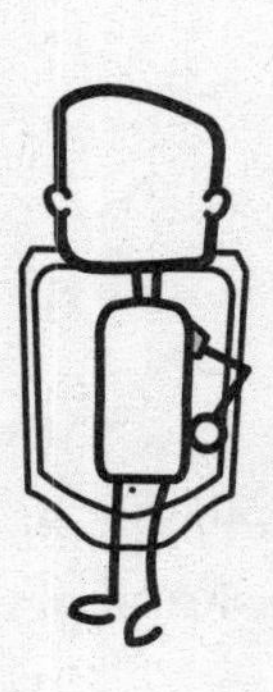
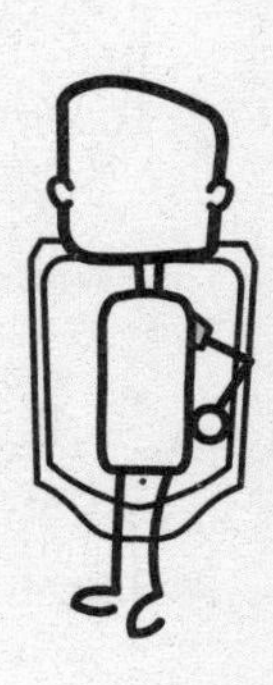

First, there's the obvious appeal of the two urinals at both ends of the row, because a man is guaranteed one side free of other urinators. If the bank of urinals is empty when a man arrives to the washroom, then he should probably choose a urinal at one end of the row or the other. But which end is best?

Did You Know . . . Historically, there have been cultures and moments in time when men sat to urinate and women stood. We have decided to use good judgment and refrain from including an illustration of this . . . but we did think about it.

Let's take the example of a bathroom with five urinals. If you're male, great. Proceed. If you're female, imagine you're a man. You enter the room and will probably take the closest urinal, the one at the end nearest you. How many more men will have to enter before some guy is forced to stand next to someone? If the next man chooses the middle urinal, and then the third man picks the urinal at the far end, it has taken three people to occupy all the spaces that have privacy on both sides.

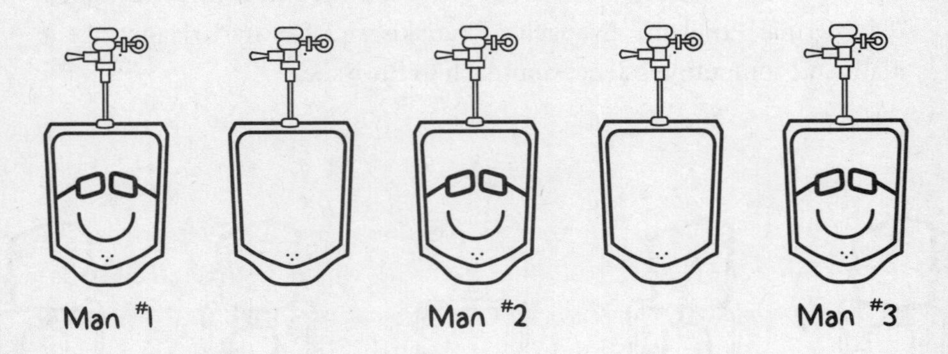

But sometimes the second guy to arrive might choose urinal #4; after all, it has an empty urinal on each side. That choice messes things up, because after he picks the fourth urinal, there are no private stalls left.

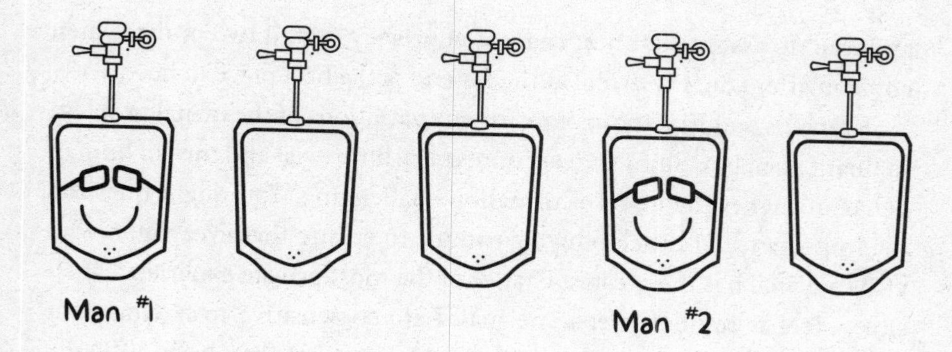

So, clearly the capacity for privacy depends on the second man's choice. As the first person in, though, you can ensure that there are three private stalls if, instead of choosing urinal #1, you choose #3 in the middle. That way the next two people will automatically occupy one end urinal and then the other one.

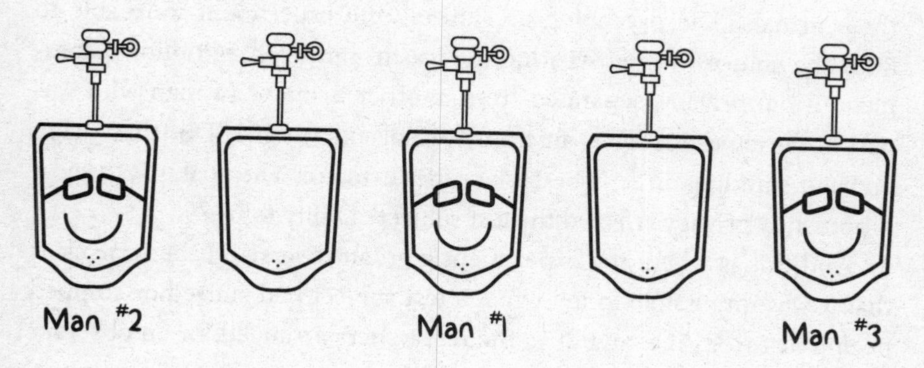

Even in simple examples like this, there are other factors to consider. For instance, Kranakis and Krizanc played around with the "Lazy Filling" strategy to see how this would affect outcome. The Lazy Filling strategy refers to men's propensity to choose the urinal closest to them. If all stalls are empty, then the man will choose to be lazy and walk to the nearest stall. When there are no longer any urinals with vacancies on both sides, the next man entering the bathroom will revert to laziness and take the urinal closest to him. Because every man's bathroom goal is to find a fully private space, it turns out that if you're the first man in the urinal and want to assure your best chances at privacy even if two or three men come in after you, the urinal at the far end is the best place to head.

Kranakis and Krizanc explored many variations of the urinal problem in their paper, but taking into account every little twist and turn of human behavior makes the math of urination challenging. You might question the importance of developing algorithms to ensure that every man who visits a urinal has the greatest chance of the most private experience possible. Here is my defense: some males are so sensitive to having their personal space invaded when they need to pee that they have difficulty starting the process at all. That's the extreme, but it appears that the desire for space, as close to privacy as you get standing at a bank of urinals, is widespread among men.

In fact, there's data to back up this claim. A controversial experiment in the 1970s had as its setting a men's washroom where there were only three urinals. The psychologists running the experiment were able to limit the choices of men visiting the washroom to three options: complete urinal privacy, separated from another urinator (a man who was part of the experiment) by an empty urinal stall (with a "Don't Use" sign on it) or standing directly beside another urinator. The goal was to see if diminished privacy affected the test subject's ability to pee.

And this is where the experiment got controversial. It became clear that it was impossible to tell when a test subject had started or stopped peeing just from the sound, so the researchers arranged for an observer to sit in a toilet stall next to the urinals with books at his feet. There was a periscope in one of the books aimed at the urinals so the observer could time "start" and "stop" accurately.

The results of this bizarre experiment were as expected: the closer a test subject was standing to the man pretending to be urinating, the more time passed before the test subject started peeing—8.4 seconds, on average. But when the bathroom was empty, test subjects peed much faster: the average time was 4.9 seconds.

The ethical issues raised by spying on people as they stood at a urinal didn't go unnoticed. It's unlikely that another study involving a man with a periscope watching other guys pee will ever be replicated. But the data from the experiment fit with other well-known observations about urinal behavior. For instance, if two male friends enter a bathroom at the same time and stand at adjacent urinals, etiquette demands they keep their eyes riveted straight ahead or down, rather than make eye contact with their friend, even if they are mid-conversation. It's all about preserving the sense of personal space, even when there isn't any.

How much do people pee in pools?

If you're one of those people who is hesitant to use the urinal in a public washroom, there's always the swimming pool or hot tub, right? Even though there's a taboo against doing that, in a 2012 survey, 19 percent admitted having done it, so you know it happens! But a percentage like that doesn't give you a clear sense of just how much urine there is in your neighborhood pool. Well, now we know.

In a study published in early 2017, University of Alberta researchers used the artificial sweetener acesulfame K (or acesulfame potassium) as a tracer for urine. It's found in products like ice cream, jam, jelly, frozen desserts, soda pop, fruit juices, toothpaste, mouthwash, and many others.

It's two hundred times sweeter than common sugar (sucrose) and is often used in conjunction with other, better-known sweeteners like sucralose or aspartame.

The beauty of acesulfame as a urine tracer is that most people consume it, our bodies don't alter it chemically as it passes through us, everyone who consumes it pees it out and it's robust enough to stick around in pool water and be detected.

Did You Know . . . That distinctive smell you associate with public swimming pools and think is chlorine? Not always. It's more often the odor of trichloramine, which forms when nitrogen in urine reacts with chlorine. Yuck!

In the study the team collected samples from two swimming pools in different Canadian cities, one with a capacity of 110,000 gallons, the other 220,000 gallons. (Olympic swimming pools are about 660,000 gallons.) The concentrations of acesulfame in the swimming pools were roughly ten times the concentrations in tap water from the two cities. The only conceivable source for this dramatic increase in concentration was urine. That meant 7.9 gallons (30 liters) of urine in one pool and 19.8 gallons (75 liters) in the other—and 19.8 gallons would fill twenty large milk jugs. That's a lot of urine!

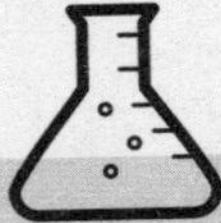

Science Fact! *There is no chemical that will turn color when exposed to urine in swimming pool water. The difficulty is identifying a chemical that would react only to pee and not other chemicals. The other roadblock is human nature: it wouldn't be difficult to imagine someone peeing in the water, then sloshing around and blaming it on someone else.*

That alone has a significant "Ewww!" factor, but there's also a health implication. While urine isn't sterile, it is not generally a risk for infection, so it isn't in and of itself a problem, but the chlorine and sweat in the pool can react with it to form what are called disinfection by-products. Those can irritate the eyes and cause breathing problems, even asthma.

It should be said that respiratory problems are most often encountered by those who spend a lot of time in pools, and most swimmers like that cheerfully acknowledge that serious swimmers pee in the pool and think nothing of it. But now you know.

What is the Turing test?

Alan Turing was a brilliant English mathematician, code breaker, and computer scientist who deciphered the German Enigma code during World War II, but he is probably best known for creating the Turing test, a way of deciding if a machine is intelligent. As such, he could be considered the father of AI, artificial intelligence.

The test, as Turing originally designed it, was straight-forward, even though his goal was profound: "I propose to consider the question, 'Can machines think?'" It took only a few more sentences before he realized that such a straightforward goal would run into problems, given the commonplace and ambiguous definitions of both "machine" and "think." To solve this complicated question, he picked a relatively simple approach: he chose to create a variation of the party game called the "imitation game."

Imagine this situation: a man and a woman sit in one room, a judge in another. The judge asks questions of both (in print, preferably typewritten) and tries to figure out, based on each person's answer, which is the woman. That's the party game. Turing tweaked that idea by substituting a computer for one of the people. The judge had to decide from the answers which was the human and which was the computer. Turing noted that the answers didn't have to be correct, only that they resembled an answer that a human might give.

Turing refined the game by saying the computer would pass the test if the human judge could do no better than accurately identify either human or computer more than 70 percent of the time, after a five-minute conversation. Turing also allowed space for future development by asking whether, if the computers of his age failed the test, "imaginable" computers could win the game. He thought the test would be won by the year 2000.

In the paper he wrote describing the test, Turing raised a number of possible objections to it, then took out each one like someone shooting ducks at an arcade. When people argued that God gives souls to humans, but not machines, Turing said, "I am not very impressed with theological arguments." And when some people said that intelligent machines were too scary a thought, he said, "I do not think this argument is sufficiently substantial." He also dismissed arguments that machines have limits, that machines cannot write poems and therefore can't think and that machines can't originate anything.

Did You Know . . . The 2014 movie *The Imitation Game* provided a glimpse into the mind and life of Alan Turing. Turing was a gay man at a time when gay relationships were considered a criminal act. He was outed in 1952, tried, and convicted. He avoided jail by agreeing to chemical castration, but died two years later from cyanide poisoning at the age of forty-two. It was assumed to be suicide, although there have been suggestions that it was accidental. Regardless, as attitudes modernized, Queen Elizabeth granted Turing a posthumous pardon in 2013.

The original version of the Turing test has yet to be officially passed, but several people claim to have beaten it. In the mid-1960s, Joseph Weizenbaum created a program called ELIZA, which was designed to mimic the conversational efforts of a style of psychotherapy called "person-centered" or "Rogerian" therapy, after the inventor Carl Rogers. The job of a Rogerian therapist is to encourage patients to reveal themselves through questions. So comments such as "How does that make you feel?" and "Does that surprise you?" are legitimate but don't really require the machine to do a lot of thinking. Weizenbaum claimed victory, although his claim has been disputed ever since.

Since then, varieties of the test have sprung up. One of my favorites is the Chinese room. Like everything in this area of research, it's contentious, but is a neat example of the kind of thinking surrounding ideas not just of machine intelligence but consciousness in general.

In this scenario, a person who knows no Chinese whatsoever is locked in a room full of information about Chinese symbols. There is also a book with directions about how to arrange these symbols in the proper order. Chinese speakers outside the room pass questions in Chinese into the room, and the person inside—without knowing that these are questions—consults the book of instructions, puts together a set of symbols, and passes them outside—again, not even realizing that the set is the answer to a question. As far as the Chinese speakers outside are concerned, there is someone (or something) inside the box that is passing the Turing test. But since the person inside knows absolutely nothing about

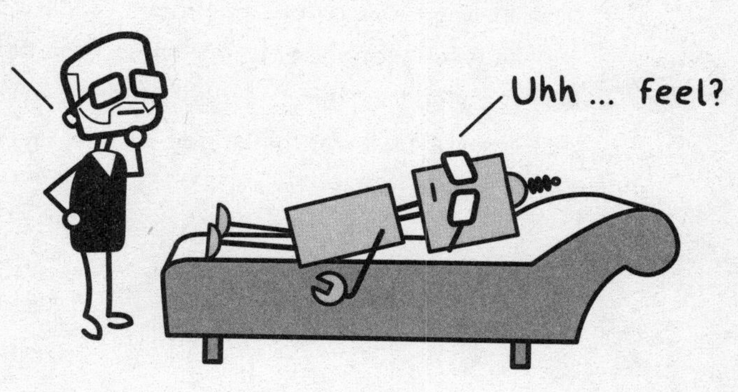

the Chinese language, they can't be said to be passing any sort of test of intelligence.

Two more of my favorites are the reverse Turing test, in which a computer has to tell if it's conversing with a person or another computer, and the total Turing test, where the computer being tested also must see and handle objects—artificial intelligence plus machine vision plus robotics.

Whether you know it or not, you've likely played a modified version of the reverse Turing test yourself. Any time you go to a website that requires you to type a set of distorted letters and numbers in a box, you are trying to demonstrate to the computer monitoring you that you are human. It's called CAPTCHA: Completely Automated Public Turing Test to Tell Computers and Humans Apart. Computers aren't nearly as good as humans at deciphering the CAPTCHA letters and numbers. Humans, on the other hand, are so good at making sense of unclear images that we can see the face of the Virgin Mary in a grilled cheese sandwich!

Did You Know . . . The Loebner Prize, created by Dr. Hugh Loebner, offers $100,000 and a gold medal of Loebner himself to the first entrant who passes the Turing test. It was started in 1991, but even with the progress in computing since then, it has yet to be won. Smaller prizes are given out to what you might call "best in show" each year, but none have reached the bar Turing set.

While writing this, I chatted with the chatbot deemed the best in the most recent Loebner competition, and it seems we still have a long way to go: when I mentioned dinner (no, I wasn't asking for a date!), it went off on a tangent about Charles Dickens. Maybe it was thinking of "Please, sir, I want some more" from his novel *Oliver Twist*.

Will machines ever have feelings?

"Stop, Dave . . . Will you stop, Dave . . . Stop Dave. I'm afraid . . . I'm afraid, Dave . . . Dave . . . my mind is going . . ."

That's HAL, the computer in Arthur C. Clarke's (and Stanley Kubrick's) *2001: A Space Odyssey*. HAL, a machine with feelings, agonizes as astronaut Dave begins to take him apart. He's reacting with emotion—with terror, in fact. Without emotion, HAL would have been nothing more than a mindless number-crunching machine.

He's fictional, though. And while we're nowhere near being able to build a machine like HAL, could we one day? And if we did, would that mean the machine with feelings would be more intelligent than all those without?

Here's some food for thought. In 1848 a man named Phineas Gage was construction foreman on a crew building the Rutland & Burlington Railroad in Vermont—routine work, putting blasting powder into a hole.

But on September 13, everything went sideways: the powder exploded, and a rod drove right through Gage's skull and landed 65 feet (20 meters) away. The rod took out his left eye and much of the left frontal lobe of his brain.

Incredibly, Gage survived the accident and lived for many years afterward. But people who had known him before testified that "good guy" Gage was gone. No longer friendly, even-tempered, and reliable, he became a bad decision maker and a sociopathic drifter. The terrible wound to his brain had transformed him.

More recently, American neurologist Antonio Damasio described a patient of his, Elliot, whom he referred to as a "modern-day Gage." Elliot had a brain tumor surgically removed, and parts of both his left and right frontal lobes were removed as part of the operation. Afterward, Elliot suddenly started making one bad decision after another. By the time Damasio saw him, he had lost virtually everything, including family and wealth. Oddly, Elliot passed myriad psychological tests, but he confessed that in many situations in which he used to have feelings, he no longer felt anything. Damasio became convinced that Elliot, like Gage before him, had lost an essential component of sound decision making: emotions.

Even lab studies have shown that people in happy moods are better able to focus on all aspects of a drawing and later reproduce it than sad people are. The happy people focus on the whole forest and can re-create the full picture; the sad ones focus on specific trees and therefore literally and figuratively fail to see the bigger picture. That suggests that a robot with "moods" might make better decisions.

But can we prove that emotionally intelligent machines might be more capable than those without feelings? And if so, how? The challenge is that human feelings are the product of millions of years of evolution, and the circuits of brain cells that produce them are very poorly understood—they're so complex. So right now, anyway, imitating them is out of the question.

Taking baby steps first, how about creating a computer that could at least recognize human emotions? Researchers in China have shown that a computer can identify positive, neutral, and negative emotions from the brain waves of humans watching film clips. Then there's the French/Japanese store robot, Pepper. It can recognize four emotions: happiness, joy, sadness, and anger. Pepper then responds in a way that might encourage shoppers to buy more. Still, these robots and computers are responding automatically; they have no idea what people are actually feeling. Even if we build computers that are able to interpret any human emotion, they still have to be able to respond appropriately.

The emotions you experience—your feelings—are inaccessible to anyone unless you express them. You might look angry to me, but that doesn't mean you actually are angry, and I have no way of probing your consciousness, that inner world of thoughts, dreams, and feelings.

Consciousness is a scientific mystery: not only do we not know how a brain generates consciousness, we don't know which living things have it. Crows and ravens are capable of insightful actions, but are they conscious? If dogs and cats are self-aware, that's evidence that a human-sized brain wouldn't be a prerequisite for a robot. Some scientists have argued that once a brain reaches a certain level of sophistication, consciousness might spontaneously emerge. Would that be true of a complex assemblage of microprocessors as well?

Did You Know . . . IBM is trying to replicate the interconnectedness of the human brain. TrueNorth is an IBM computer chip that simulates neurons—16 million of them, which in turn make 4 billion connections. That's impressive, although 16 million can't compete with the 86 billion in the human brain.

If robots were able to develop emotions, would they even be like ours? Maybe they'd have "machine feelings": no "heart," just a battery. A machine with feelings might be a more effective machine, but it might also be much less predictable. And if all those sensitive but smart robot types decide they don't need us around anymore . . . then what?

Did You Know . . . Scientists in Germany showed people a computer-animated scene of a man and a woman discussing their feelings about the hot weather, their lack of spare time, and the woman's annoyance at having been stood up by a friend. Onlookers were told either that the voices were human or computer-generated. Those who were told the conversations were computer-generated and unscripted thought the scene was eerie. Those who were told these were human interactions did not. What does this mean? Apparently we feel that there's something disturbing about a computer that is thinking entirely on its own. But we might just have to get used to that.

Are we living in a computer simulation?

Before you let that question stop you dead, consider this: the Greek philosopher Plato told of people chained in a cave who saw shadows cast on the wall. The shadows belonged to puppets and the voices the people heard belonged to the puppet masters, not the shadows. But as far as the people knew, those shadows and voices were "reality." Perhaps we are no different, limited by our senses and our brains to some incomplete version of what really exists.

Today, if you've experienced high-quality virtual reality, you know that an artificial environment can be incredibly compelling. I've been in a VR version of a high-rise office where there was a balcony with no railing. When I was invited to step off into "thin air," I couldn't force myself to do it, even though I knew I was standing on a factory floor. The "reality" presented to me was simply too convincing.

And we have computer-generated alternative societies like Second Life, where you can have a perfectly normal conversation with an avatar whom you assume represents a real human somewhere, although you have no evidence for that.

Now consider the future, when computers will be immeasurably more powerful than they are today. VR and Second Life will seem like crude crayon drawings in comparison. Ray Kurzweil, inventor, futurist, and engineer, has argued that the time when computers will be more intelligent than humans will come in the next ten to fifteen years. In 2030, a computer costing a thousand US dollars will be one thousand times more powerful than the human brain.

Then imagine a time when computers are powerful enough to simulate the entire history of the earth and all the people who have ever lived on it. And they're common! Someone populates it with self-aware automatons and lets it run.

Each individual in this computer scenario would have the complete consciousness of any human today. In fact, each might actually be any human today. If you were one of those automatons, how would you know? You wouldn't. Everything in your personal past—every historical event and everything that hasn't yet happened to you but will—would simply be a computer program running its software. And you would have no idea.

You would be just another character in an amped-up version of the Sims, living your life in the equivalent of SimCity. We know—or at least are pretty sure!—that today's Sims aren't self-aware, but with much greater computer capacity, who can say they wouldn't be? When people like inventor Elon Musk claim that "either we're going to create simulations that are indistinguishable from reality, or civilization will cease to exist," the idea is worth a closer look.

There are issues. First, will we survive to the point where we'd be capable of creating what philosopher Nick Bostrom likes to call "ancestor simulations"? It's conceivable that some sort of catastrophe might interrupt our progress toward that point, perhaps permanently. (See "Are we alone in the universe or are aliens out there?" on page 219.)

If we do achieve the kind of computing power necessary, would any citizen in that computer-enhanced world bother to do something like this? We have no way of knowing, but it would have the lure of being the closest thing to playing God that we've ever seen. (Of course, there is the idea that such a person would, in many ways, be playing the role of the many gods that have been worshipped throughout history. Gods have ultimate power over the civilizations that believe in them; computer operators would have ultimate power over their simulations.)

Science Fiction or Fact! *It's true that real-life simulators don't bother to simulate the detail of a scene if it isn't important. They take shortcuts: maybe the people in the simulation we live in whom we think are conversing are really just saying those stock phrases uttered by movie extras to mimic conversation: "Peas and carrots, watermelon cantaloupe, or rhubarb, rhubarb." Listen closely next time!*

The programming of such a simulation would be a significant stumbling block. Some 86 billion neurons in the human brain, together with an even greater number of cells whose role is largely unknown, produce vivid ideas and sensations. Cells are flesh and blood; thoughts and ideas are not. How does one make the other? To give a simulation's inhabitants fully human qualities, human consciousness would have to be replicated.

If we were living in a simulation, would we know? Any computer that can simulate an entire civilization and its inhabitants would likely have safeguards against detection. But Cambridge mathematician and cosmologist John Barrow wonders if the simulation simply wouldn't be accurate enough. When Disney creates an image of light reflecting off the water, it doesn't get it absolutely right; it's just "good enough, as long as no one looks too closely." And as our knowledge increases, simulation software would have to be continually updated in the same way smartphone apps are. Then observant physicists might notice tiny mysterious changes that would alert them that something is amiss.

A usual question is: "If we're living in a simulation, how should we behave?" The straight answer is: "We have to encourage whoever's running the simulation to keep it going. If they get tired of it, it's game over!"

But how do we do that? Here are suggested tactics:

- Live for today (because we don't know when they might pull the plug!).
- Be entertaining (because that will hold their interest and prevent the plug from being pulled).
- Try to figure out what qualities they desire in their simulated people (in case they're the kind who want to play God).
- Try to encourage the existence of famous and fascinating people (we all love stories and want to be entertained ourselves).

There are people who are doing those very things. But not everyone buys it. Of simulations like this, Elon Musk said, "My brother and I agreed that we would ban such conversations if we were ever in a hot tub."

History Mystery

Why were *Tyrannosaurus rex*'s arms so short?

Tyrannosaurus rex was not the biggest carnivorous dinosaur ever. As far as we know, that distinction goes to the peculiar swamp-dwelling *Spinosaurus*.

T. rex isn't even the biggest and fiercest of our imaginations—that's *Jurassic World*'s CGI colossus, *Indominus rex*. But who cares? It's still *T. rex*, and whether it occupies the all-time heavyweight throne or not, it was the ruler of the Cretaceous period. Besides, I challenge you to name another dinosaur that inspired the name of a rock band.

A typical *T. rex* could reach 40 feet (12 meters) in length, measure 13 feet (4 meters) at the hips, and weigh nearly 7 ½ tons (7 tonnes). To grow to those dimensions, teenage *T. rex* would go on a weight-gaining spree, putting on 1,300 pounds (600 kilograms) a year from ages fourteen to eighteen. It'd go on to live into its mid-twenties. Even at that size, it could manage speeds of roughly 18 miles an hour (8 meters a second), making it a fierce predator. Its astounding bite force was ten times as powerful as that of an alligator, making it the most powerful ever seen on Earth. But there is one feature of the *T. rex* that has always puzzled experts: its tiny forelimbs.

On a full-sized *T. rex*, the forearms were only about three feet (one meter) long. There are basketball players with a reach like that. When he described the first skeleton of *T. rex* in 1905, the American paleontologist Henry Fairfield Osborn—having recovered only the dinosaur's upper arm bone, the humerus—imaginatively reconstructed the beast with a long, three-fingered forearm. The first complete set of *T. rex* forelimb bones wasn't discovered until 1989, and those bones confirmed what many had suspected from studying related species: the *T. rex* had absurdly short arms—too short even to reach its mouth.

The immediate question, then, was what were those tiny arms good for? Some dismissed them as vestigial, envisioning a progressive shrinking from the full-length prey-catching arms of smaller carnivorous ancestors to the stunted forelimbs of the *T. rex*. The problem there is that however short they might have been, the *T. rex*'s forearms were heavily muscled. Inferring power from fossil traces is one thing, but it's also possible to create mathematical models of the *T. rex* forearms, plug in known values, and calculate what they were capable of. Both approaches arrive at the same result: the forearms were powerful and fast.

In fact, to say they were heavily muscled would be an understatement. Judging from the thickness of the arm bones—three times the size of an average person's—and the huge areas on them for the attachment of muscles and tendons, we can conclude that the arm was at least the thickness of a human thigh. The biceps attached farther down on the forearm than ours, giving the *T. rex*'s arm more leverage, and they had help from the surrounding muscles, especially the huge shoulders. With all that power behind them, each arm alone could curl close to 450 pounds (200 kilograms)! If evolution was gradually marginalizing the arms, though, it wouldn't make sense to retain that kind of musculature. The arms must have had a role to play.

TRY THIS: Straighten your arm in front of you, then touch your shoulder. That one movement takes your elbow through 165 degrees. Because of all the muscle on its forearm, the *T. rex* could manage only 45 degrees, with very little side-to-side movement. It just wasn't built to do anything more than simple flexing.

Osborn speculated that the muscle on the arms might have been involved in sex—but don't forget, he didn't know how incredibly short those arms were. So what exactly did these powerful but apparently undersized pieces of machinery do? Some scientists used the shortness of the arms as evidence that the *T. rex* didn't run very fast. If a *T. rex* had tripped, its little arms wouldn't have been able to break its descent, and such a fall for a beast that size would likely have been fatal: given its weight, it would have slammed into the ground with six times the force of gravity and broken bones in the process. But this theory was questioned by referencing other animals. Giraffes can sprint 30 miles per hour (50 kilometers an hour) and would likely break a leg if they fell. Ostriches move very fast, too. Neither animal has arms to break a fall, and so there's really no reason to think *T. rex* was any different.

My favorite explanation was offered by paleontologist Barney Newman in 1970. Newman envisioned a *T. rex* "in a position of rest," lying facedown with its jaw on the ground. How would it get up? The power would have to come from the giant hind legs, but if they pushed with no counterforce, the animal would skid forward on its belly. Instead, Newman suggested, the first thing the dinosaur would do is dig its front claws into the ground and do a push-up. That would allow it to rock backward and upward, eventually standing on its hind legs.

Newman's idea hasn't exactly been shot down, but it hasn't prevailed, either. Much more emphasis these days is placed on the notion that the *T. rex*'s mini-arms were somehow useful in feeding. There is evidence that they were used in hazardous situations, because many of the surviving fossil *T. rex* arm bones have been chipped, cracked, and broken, apparently by excessive forces. If all those arms were used for was propping the animal up, it's hard to see how they would incur such damage. At the same time, it's also challenging to imagine a *T. rex* using those tiny, bulging arms to help position and subdue a gargantuan *Triceratops* so that its giant jaws could make the kill.

There's always been some controversy about whether the *T. rex* was a mighty hunter or simply an oversized scavenger, the multi-ton buzzard of the dinosaur age. Consensus seems to be building in support of it being a hunter.

In fact, according to Canadian paleontologist Phil Currie, *T. rex* was a pack hunter, like the wolf. The best evidence of this is the fact that the *T. rex* had a brain three times the size of other dinosaurs—pack hunting requires smarts to coordinate the activities of several animals simultaneously. A set of tracks found in British Columbia of several tyrannosaurs (either *T. rex* itself or that of a relative) walking together further supports the theory.

If *T. rex* was smart, though, there were limits to its intelligence. There is one famous fossil of a *Triceratops*, the three-horned dinosaur with the huge frilled shield on its head, where one horn has been broken off and part of the frill has been chewed. *T. rex* is the only possible candidate for the attack. Is it intelligent to attack such an animal—which would have been the size of an elephant—from the front, where all of its weapons and armor were? Well, no, it isn't.

You'd think that with two decades of the *Jurassic Park* brand under our belts, we'd have a fairly accurate depiction of *T. rex* in our minds. Especially in kids, who devour anything dinosaurian. Apparently not. A trio of educators at Cornell University tested both college and school-age students for their understanding of *T. rex*'s posture. When *T. rex* was first unveiled to the public, in 1905, it was depicted standing upright, dragging its immense tail along the ground (see Exhibit A).

EXHIBIT A

That was no more than a guess—and it was wrong. When scientists started to dig deeper, they realized that tail-dragging made no sense, not to mention that none of the fossil trackways of carnivorous dinosaurs like *T. rex* showed any signs of a dragging tail. In fact, an upright posture would likely have damaged several of *T. rex*'s joints, including the hips. Instead, think of *T. rex* as a horizontal animal, its head extending out front and its tail straight out behind. The animal likely swaggered a little as it walked, swinging its tail from side to side to balance the huge weight shifts caused as each giant hind leg stepped forward (see Exhibit B).

EXHIBIT B

To its credit, *Jurassic Park* portrayed the dinosaur the right way (although *T. rex* existed in the Cretaceous period, not the Jurassic), and that was in 1993. A full generation since then has had a chance to adopt the correct posture of *T. rex* for its shared depictions of the animal. But the Cornell experiments suggested the exact opposite had happened. In that study, 111 students were asked to sketch a standing *T. rex*; the average angle of the spine relative to the ground depicted was more than forty-five degrees, closer to the 1905 original and far from today's interpretation, which would be closer to zero degrees!

It seems unbelievable that a dramatic change in the posture of the world's most famous dinosaur has failed to penetrate the minds of its biggest fans. But there it is. I blame Barney.

Acknowledgments

There I was, wondering if I would ever write another book, when my old friend Kevin Hanson sprung one on me. Then another. And here they are.

Nita Pronovost, with Brendan May, shepherded these manuscripts from beginning to end, appearing benevolent but I swear I glimpsed a whip behind their backs. The Simon & Schuster publicity team, especially Catherine Whiteside, has worked hard to sell these books.

My agent, Jackie Kaiser, and everyone at Westwood Creative Artists, have made the whole process run smoothly—writing with an agent is a whole lot better than writing without one.

I had help on these books from two fantastic researchers, both of whom I know from the Banff Science Communications Program at the Banff Centre. Niki Wilson, an accomplished science writer on her own, and Joanne O'Meara, a physics prof at the University of Guelph and a great communicator, too, took on the job of scouring the scientific literature on biology (Niki) and physics (Joanne).

Finally, I'm lucky to have a set of inquisitive friends who will tolerate my talking (babbling) about this or that scientific. They include Wendy Tilby, Amanda Forbis, Dave Nicholls, Denise Jakal, Trevor Day, Garth Kennedy, Ben Jackson, Steve Dodd, Josip Vulic, and Penny Park. And the Ingrams and the Westburys.

Mary Anne, of course, serves as yet another set of eyes and—way more important—ongoing support.

Thanks to all.

About the Author

© RICHARD SIEMENS

JAY INGRAM has written seventeen books, including the bestselling first book of this series, *The Science of Why*. He was the host of Discovery Channel Canada's *Daily Planet* from the first episode until June 2011. Before joining Discovery, Ingram hosted CBC Radio's national science show, *Quirks & Quarks*. He has received the Sandford Fleming Medal from the Royal Canadian Institute, the Royal Society of Canada's McNeil Medal for the Public Awareness of Science, and the Michael Smith Award for Science Promotion from the Natural Sciences and Engineering Research Council of Canada. He is a distinguished alumnus of the University of Alberta, has received five honorary doctorates, and is a Member of the Order of Canada. Visit Jay at JayIngram.ca.

@jayingram